101 FACTS

YOU DIDN'T KNOW ABOUT

SPACE

101 FACTS
YOU DIDN'T KNOW ABOUT
SPACE

MARK THOMPSON

WHITE OWL

AN IMPRINT OF PEN & SWORD BOOKS LTD.
YORKSHIRE – PHILADELPHIA

First published in Great Britain in 2020 by White Owl
An imprint of
Pen & Sword Books Ltd
Yorkshire - Philadelphia

ISBN 9781526744579

Design by Paul Wilkinson

Printed and bound in India by Replika Press Pvt. Ltd.

Pen & Sword Books Ltd incorporates the Imprints of Pen & Sword Books Archaeology, Atlas, Aviation, Battleground, Discovery, Family History, History, Maritime, Military, Naval, Politics, Railways, Select, Transport, True Crime, Fiction, Frontline Books, Leo Cooper, Praetorian Press, Seaforth Publishing, Wharncliffe and White Owl.

For a complete list of Pen & Sword titles please contact:

PEN & SWORD BOOKS LIMITED
47 Church Street, Barnsley, South Yorkshire, S70 2AS, England
E-mail: enquiries@pen-and-sword.co.uk
Website: www.pen-and-sword.co.uk

or

PEN AND SWORD BOOKS
1950 Lawrence Rd, Havertown, PA 19083, USA
E-mail: Uspen-and-sword@casematepublishers.com
Website: www.penandswordbooks.com

CONTENTS

FACT 1: A compost heap generates as much energy as the Sun!............................8

FACT 2: Venus smells like rotten eggs 10

FACT 3: You are made of stardust 12

FACT 4: Footprints on the Moon will last for millions of years 14

FACT 5: Oceans of liquid diamond may exist on Neptune 16

FACT 6: Astronauts have feet as soft as babies 18

FACT 7: Space is completely silent 20

FACT 8: Hot stars are blue, cold stars are red... our taps are wrong! 22

FACT 9: Some stars spin 700 times per second 24

FACT 10: Venus and Uranus are upside down 26

FACT 11: Saturn is not the only planet with rings 28

FACT 12: Dark sunspots are as bright as the full Moon 30

FACT 13: Metal sticks together in space ... 32

FACT 14: Astronauts cannot burp in space 34

FACT 15: The days really are getting longer 36

FACT 16: Neptune is home to the fastest winds in the Solar System....................... 38

FACT 17: There is no dark side to the Moon 40

FACT 18: Black holes are not actually holes! 42

FACT 19: Clusters of galaxies are used as gigantic cosmic telescopes.......................**44**

FACT 20: Mercury is the fastest planet...**46**

FACT 21: On Mars you need flip flops and a hat...**48**

FACT 22: The moons of Jupiter told us the speed of light...................................**50**

FACT 23: Earth is travelling through space at 225 km per second.........................**52**

FACT 24: The light from stars tell astronomers what the Universe is made of..**54**

FACT 25: Space agencies use planets to steer spacecraft around the Solar System..**56**

FACT 26: Shepherds keep the rings of Saturn in order.....................................**58**

FACT 27: The Universe is expanding faster now than in the past.................................**60**

FACT 28: Zero gravity makes copulation rather tricky...**62**

FACT 29: The sight of a larger than usual Moon rising is just an illusion.................**63**

FACT 30: Twinkle twinkle little star! Stars do not twinkle.......................................**64**

FACT 31: Europa, a moon of Jupiter, could harbour alien life.....................................**66**

FACT 32: Pluto is no longer a planet.......**68**

FACT 33: The first spacewalk nearly ended in tragedy.....................................**70**

FACT 34: The Andromeda galaxy is on a collision course with the Milky Way........**72**

FACT 35: A Martian volcano is almost three times the size of Mount Everest ...**74**

FACT 36: The telescope was invented by an optician.................................**76**

FACT 37: Astronomers use black and white cameras to take colour pictures..................**77**

FACT 38: There are thousands of planets beyond our Solar System............................**78**

FACT 39: There is a hurricane on Jupiter nearly three times the size of Earth **80**

FACT 40: If you fell into a black hole you would be stretched like spaghetti...........**82**

FACT 41: There are clouds of water floating in space.................................**84**

FACT 42: It is very likely that the Sun will one day swallow up the Earth....................**86**

FACT 43: Shooting stars are not stars at all...**88**

FACT 44: Inhabitants of Saturn could enjoy the northern lights too **90**

FACT 45: Comets have two tails...............**92**

FACT 46: The Moon was once part of Earth...**93**

FACT 47: To find black holes, astronomers look for the brightest objects in the night sky...**94**

FACT 48: There are three laws which govern the motion of the planets............**95**

FACT 49: The most common star is a red dwarf..**96**

FACT 50: The Kessler effect suggests one day we may get cut off from space!.........**98**

FACT 51: The first telescope mirrors used arsenic **100**

FACT 52: A gas cloud contains enough alcohol for everyone for a billion years ...**102**

FACT 53: Our eyes are rubbish colour detectors in the dark...................................**104**

FACT 54: Light takes around one million years to reach us from the core of the Sun...**105**

FACT 55: A hurricane force wind is just a breeze on Mars **106**

FACT 56: An astronaut in space would last about 15 seconds without a spacesuit ...**108**

FACT 57: Travel fast to stay young...........**110**

FACT 58: The sky has co-ordinates just like latitude and longitude.........................**111**

FACT 59: The Grand Canyon is dwarfed by the largest valley in the Solar System **112**

FACT 60: Earth is closer to the Sun in winter...**114**

FACT 61: There is water on the Moon....**116**

FACT 62: Dark matter exists but we do not know what it is **118**

FACT 63: Studying the light from galaxies tells us how fast they are moving...........**120**

FACT 64: The largest single mirror telescope has a mirror 8.2m across **122**

FACT 65: Pigeon droppings were once mistakenly identified to explain the radiation from the Big Bang....................**123**

FACT 66: Saturn's moon Mimas looks like the Death Star from Star Wars.................**124**

FACT 67: Neptune was first discovered by mathematics..**126**

FACT 68: Meteorites are not hot!.............**127**

FACT 69: The Sun looks white in space..**128**

FACT 70: Voyager 1 will reach its destination in 40,000 years........................**130**

FACT 71: We are sending signals into outer space and aliens could be listening......**132**

FACT 72: Mercury is shrinking!................134

FACT 73: The Hubble Space Telescope was launched with the wrong mirror............136

FACT 74: A veritable menagerie of animals have been sent into space...........138

FACT 75: The movement of the Earth allows astronomers to measure distances in space................139

FACT 76: Stars do not live forever..........140

FACT 77: The different shapes of galaxies relate to their evolutionary stage..........142

FACT 78: The rotation of the Sun is not uniform.................144

FACT 79: Some variable stars can tell us how far away they are.................146

FACT 80: Spacecraft docking in space is a tricky manoeuvre.................148

FACT 81: The Universe is 92 billion light years across.................150

FACT 82: In 1977 astronomers thought they had detected an alien radio signal.........151

FACT 83: Gamma ray bursts are among the most powerful explosions in the Universe.................152

FACT 84: The constellations of today will not be recognisable in a hundred thousand years.................154

FACT 85: VY Canis Majoris is the largest known star in the Universe.................156

FACT 86: Jupiter is a failed star..............158

FACT 87: Dung beetles use the Milky Way to navigate.................160

FACT 88: Jupiter is the Solar System's vacuum cleaner.................162

FACT 89: Polaris is the north pole star but in 12,000 years it will be replaced by Vega.................164

FACT 90: The night sky in a globular cluster would be glittering with thousands of bright stars.................166

FACT 91: Eratosthenes measured the circumference of Earth in 240bc............167

FACT 92: A star called Lucy is a large cosmic diamond.................168

FACT 93: Lunar eclipses cause massive temperature drops on the Moon...........169

FACT 94: A teaspoon of neutron star material weighs 10 million tonnes.........170

FACT 95: The Pistol Star is 10 million times brighter than the Sun.................172

FACT 96: The atmosphere of the Sun is hotter than its visible surface.................174

FACT 97: Some rocket engines produce enough thrust to lift a sheet of A4 paper!....176

FACT 98: The Sun is a very faint star!....178

FACT 99: Astronomers have their very own tape measures.................179

FACT 100: A Martian meteorite has the fossilised building blocks of life inside it. 180

FACT 101: There are alien lakes on Saturn's moon Titan.................182

Compost heap generates as much energy as the Sun!

The Sun is at the very heart of our Solar System and on average 150 million kilometres away from us. It rises in the east and sets in the west, has been there since we were born and should be there until the day we die! To many it never seems to change, but get close up and personal with the Sun and you begin to appreciate that it is so much more than just a glowing ball of light in the sky that makes us warm on a sunny day.

The Sun is a massive ball of gas, more accurately a ball of plasma, which is an electrically charged gas. It has a visible surface which we call the photosphere, where temperatures vary between 4,500 K and 6,000 K (kelvin scale of temperature measurement) and it is here where sunspots can be seen. Directly below the photosphere is the convective zone, which is 200,000 km thick, and it is here that energy from the core of the Sun is transferred through convection (the process where warmer material is less dense and rises while cooler material is more dense and sinks). Below the convective zone is the radiative zone, which is 300,000 km thick, and where the process of radiation transfers the heat. Beneath the radiative zone is the core of the Sun and it is here that the energy is actually produced. The core of the Sun has a radius of 153,000 km and the temperatures soar to a toasty 15 million K!

The process by which the Sun creates its energy is through fusion, the joining of hydrogen molecules to create helium molecules. It is actually a little more complex than that but it takes four hydrogen molecules to fuse together to produce a helium molecule, and when they do, they give off a little bit of energy. This energy is the heat and light we experience from the Sun but the figures are vast and almost incomprehensible. Every second, the fusion process converts around 700 million tonnes of matter from hydrogen into helium, but 5 million tonnes gets converted into energy (in accordance with Einstein's famous equation $E=mc^2$).

The nuclear fusion happens in the core, which as we have seen is 306,000 km in diameter, but the Sun itself is 1.39 million km in diameter so the vast majority of energy production happens in a relatively small volume. The power output of the core is estimated to be 276 watts per cubic metre which, when compared to a typical household light bulb of 60 or 100 watts is not a great deal and is around the same as a fairly typically sized compost heap!

Surprising? Even though the power output is 276 watts per cubic metre there are LOTS of 'cubic metres' in the Sun, 15 million billion cubic metres to be exact, so if you multiply the 276 watts per cubic metre you can understand how the power output volume for volume can be less than a compost heap, but the overall output of the Sun is much greater.

The Sun as seen by the SOHO (Solar and Heliospheric Observatory) spacecraft, revealing detail in the solar chromosphere. (NASA)

FACT 2

Venus smells like rotten eggs

What does Earth smell like? It is pretty tricky to name one smell because it largely depends where you are and even on the weather. For example cities and towns smell of many different things including traffic fumes whereas a forest has that lovely 'green' smell we associate with nature. If it rains then the impact of drops of water on soil can release aerosol gasses giving rain that 'fresh' smell. Other planets have smells too but until human visitors can breathe in their air (which is highly unlikely due to their atmospheric composition) we can only guess what they smell like.

Consider Venus. Whilst Venus is the second planet from the Sun it surprises most that it is hotter than Mercury, which is closer. The conditions on Venus are really quite hostile thanks to the greenhouse effect. You will have heard that phrase before and it originates from the way certain gasses cause a planet to warm up. The warmth you feel on a sunny day isn't solely because energy from the Sun hits your body and warms you up, instead, energy from the Sun traverses through our atmosphere, warms the ground, which then re-radiates the energy back into the atmosphere, which makes the lower atmosphere where we live nice and warm. A lot of the re-radiated energy then escapes out into space, moderating the temperature. The presence of greenhouse gasses can stop the energy radiating out into space just like glass stops heat escaping from a greenhouse.

Venus has suffered from the greenhouse effect for millions of years due to its proximity to the Sun and to significant volumes of carbon dioxide, a greenhouse gas, being released into its atmosphere. The Venusian atmosphere is almost entirely carbon dioxide and, with the release of sulphuric acid from volcanoes and atmospheric chemical reactions it even rains sulphuric acid from the thick dense clouds. The temperature in the atmosphere means the raindrops evaporate before they hit the ground. Thanks to spacecraft that have visited Venus such as Vanera and Mariner we now have a very good model of its atmosphere and can infer that the presence of sulphuric acid and hydrogen sulphide will cause the spacecraft of anyone brave enough (or stupid enough) to venture to Venus to fill with the rather pungent smell of rotten eggs as you prepare for landing!

Visible light image of Venus
showing the tops of its
thick dense atmosphere.
(NASA)

FACT 3

You are made of stardust

Look around you, what can you see? If you are at home then you might see other members of your family, a sofa and perhaps a TV, or if you are sat in a coffee shop you might see a street scene with cars and buses. Everything you can see is made up of a multitude of different atoms from iron to carbon, but when the Universe formed over 13 billion years ago the first atoms were mostly hydrogen and helium. Somehow, something happened, some physical process turned hydrogen and helium into the stuff we see in the Universe today.

The processes that cause this almost magical change occur deep inside the stars. After the Universe formed, bringing with it a seething, boiling soup of energy, eventually matter in the form of hydrogen and helium formed. In time, this matter in the form of gas started to accumulate into localised regions until the pressures inside became so high that nuclear fusion started to take place. Fusion is a physical process whereby atoms join or fuse together to form another type of atom. The onset of fusion marks the birth of a new star and this first generation of stars would have been almost entirely composed of hydrogen and helium.

For millions of years, these new stars would sit, quietly fusing hydrogen to helium in their core and as a byproduct emitting energy in the form of heat and light (among other emissions). Eventually the star would end up with a helium rich core.

Skipping forward a few billion years and the stars we see in the night sky start their lives in much the same way and can remain stable like this for billions of years. The stability is a beautiful balance between the force of gravity trying to collapse the star and the thermonuclear pressure from fusion trying to rip the star apart.

Once a star has a helium core, the temperature and pressure in the core is not high enough for helium fusion to occur, leading to a decrease in the thermonuclear pressure. For a short while, gravity starts winning and the core contracts, leading to an increase in core temperature and pressure and the start of helium fusion. The fusion of helium creates carbon in the core of the ageing star and for stars like our Sun it is the end of the road and the star will soon die. More massive stars live longer, and eventually carbon fuses into oxygen, then into silicon and finally iron. Even for the most massive stars in the Universe, this is as far as they can go. The thermonuclear pressure drops and the star collapses in an instant under the immense force of gravity leading to an explosion.

The process of synthesising elements inside stars like this is known as nucleosynthesis, and as the stars die the elements created inside them are scattered throughout the Galaxy. With the new elements distributed around space, they eventually get caught up in new regions of star formation and, as the stars form, the heavier elements can now play their part in the formation of planets and ultimately life.

It is true that every atom in your body has at some point, been through the core of a star.

Star cluster Westerlund 1 is 15,000 light years away and is revealed in all its glorious detail in this image from the Hubble Space Telescope. (NASA)

FACT 4

Footprints on the Moon will last for millions of years

Walk around the beach or across a field and you will see the imprint left by your feet. Come back to the same spot a few days later and you will more than likely find no evidence that you were ever there. The process that mysteriously removes your footprint is known as erosion and it can only occur on a world with a reasonably dense atmosphere like Earth. At the surface, there are about 100 billion billion molecules per cubic centimetre in the Earth's atmosphere, but on a world like the Moon there are only 100 molecules per cubic centimetre.

Erosion is a term that explains the process of the movement of surface material from one location to another. This transportation of surface material can be through movement of water, wind or even ice in glaciers. Wind and water transportation are probably the more familiar to us as they can be seen on very regular basis, with wind dislodging surface particles of soil or sand and rainfall causing water movement over the surface. Both of these require an atmosphere because without an atmosphere there will be no wind nor any rain. These are plentiful on Earth, especially in the UK, and easily erase a footprint, but on the Moon the situation is very different.

The Moon does have an atmosphere but it is so thin it is of no consequence and certainly not thick enough to experience any weather. The Earth has an atmosphere thanks to a long history of strong geological activity like plate tectonics. The movement of the surface crust leads to volcanic and other activity which feeds the Earth's atmosphere; however on the Moon there are no such processes and it is to all intents geologically dead. Any gasses that might find their way out of the lunar interior will soon be lost to space. Gravity is a significant factor in an astronomical world having an atmosphere, and the Moon, which is only 1.2% the mass of the Earth, simply does not have enough gravity to hold on to an atmosphere. The proximity to the Sun is also a factor with the constant stream of charged particles known as the solar wind pushing against any gas hugging the Moon; the weak gravity is unable to keep hold of it.

When Neil Armstrong went to the Moon back in 1969, he and Buzz Aldrin left their footprints on the surface. The lack of atmosphere means erosion on the Moon is a slow process but there is still erosion, not from the movement of surface material from wind or rain but by the solar wind itself and from space rocks known as meteorites. There is plenty of evidence of meteorites having hit the surface of the Moon from the huge dents left behind that we call craters. It is possible a big meteorite could smack into the surface and obliterate the footprints tomorrow but it is more likely that the solar wind and tiny micrometeorites will very slowly and gently wipe away the surface features. This process of erosion is much slower than the erosion we experience and it could very easily take millions if not billions of years to remove evidence that the brave astronauts of *Apollo 11* visited our astronomical neighbour.

The footprint left on the Moon from the *Apollo 11* mission. (NASA)

FACT 5

Oceans of liquid diamond may exist on Neptune

If all the water on Earth from the rivers, lakes, oceans, reservoirs, in the ground and even in the atmosphere were contained in one big drop of water, it would be 1,385 kilometres in diameter. Head to the coast here and you can enjoy a paddle in the sea. Water on Earth is pretty common: in fact liquid water covers 71% of the surface of our home planet. It is not just Earth where we find water in some form or another, its chemical signature has been found on the Moon, Mars and even on comets. As we search the Universe for planets like our own we envisage finding worlds with oceans and rivers of water like our own, but as we uncover the secrets of alien worlds they are a constant surprise to us. Even worlds closer to home, like Neptune in our own Solar System, can still surprise us.

Neptune has a diameter of 49,244 km and is 17 times more massive than the Earth. In figures that equates to a mass of 1.02×10^{26} kg, which is 102 followed by 24 zeros. Like Jupiter, Saturn and Uranus, Neptune is often said to be a world of gas, and indeed for the most part it is predominantly gaseous in nature, with hydrogen, helium, methane and carbon being present. Dive deep down into the layers and hostile conditions are to be found where pressures and temperatures have risen to extremes that have a surprising effect on the carbon found in Neptune's lower levels.

We know that diamonds are made of carbon just like the carbon found at Neptune and usually, if you heat diamonds to a high enough temperature they turn to graphite, another form of more stable carbon. Experimentation has shown that if you heat them to high temperatures and expose them to pressures in excess of 40 million times greater than Earth's average sea level pressure, then the diamonds melt to form liquid diamond. These are the conditions found deep inside the atmosphere of Neptune so it is highly likely that there are oceans of liquid diamond.

The liquid diamond oceans sound wonderful. Reduce the pressure a little and solid chunks of diamond start appearing and these chunks float, just like icebergs. A visitor would have to survive crushing pressure and extreme temperatures but if they could, then down in the depths of Neptune you might not only find oceans of liquid diamond but may even get to see these diamond icebergs floating around.

Neptune, the outermost planet in the Solar
System, where conditions may allow for
oceans of liquid diamond. (NASA)

FACT 6

Astronauts have feet as soft as babies

Strange as it may sound, when you are in space, your feet serve a completely different function. Here on Earth the force of gravity pins you to the surface and your feet provide three key functions: helping to balance you, supporting your weight and propelling you around. In space, they are useful but for a different reason, and to understand why, we need to learn about orbits and free-fall.

Imagine you were to board a rocket, get strapped to your seat and experience the immense acceleration of lift off. Once your rocket reaches 'orbit' the commander gives word that you can unlatch you harness and as you do, you gently float away from your chair: you are weightless. Well, not really without weight, instead you are in a state of free-fall and can float effortlessly around, your feet now almost redundant. To understand orbits, imagine you are back on the Earth where you pick up a small ball, let go and as you might expect, gravity pulls it back to the surface of the Earth. Now try and throw the ball, carefully watching the path it takes. You will see it follows a curved path back to the ground. Throw it harder and it still follows a curved path, shallower this time but still it lands. We know the Earth is itself roughly the shape of a ball so its surface is curved. Now imagine all the buildings, trees and mountains are gone and that the surface is smooth like a massive snooker ball. You can easily imagine that if you throw the ball hard enough then it will start curving to the surface but the surface will curve away from it so it will never land. Eventually the ball will hit you on the back of your head or continue going round. The ball is in orbit.

Spacecraft in orbit around Earth are constantly falling towards the surface but the surface is constantly falling away from them, or in other words, they are in free-fall. If you were inside a spacecraft that is in orbit then you too would be free-falling and experiencing weightlessness. It is not just astronauts and spacecraft that orbit: the Moon is in orbit around Earth and the Earth is in orbit around the Sun.

You can see now that if you are in space, in orbit and free-falling, then your feet no longer need to hold your weight, provide balance or even propel you along because you no longer need to walk anywhere. Astronauts on board the International Space Station do not wear shoes, just socks, and because the bottoms of their feet are not being used, all the rough, dried skin eventually falls off. Interestingly though, the tops of their feet start to harden and form calluses because the space station is full of bars under which astronauts hook their feet to keep them in one place and stop them floating off.

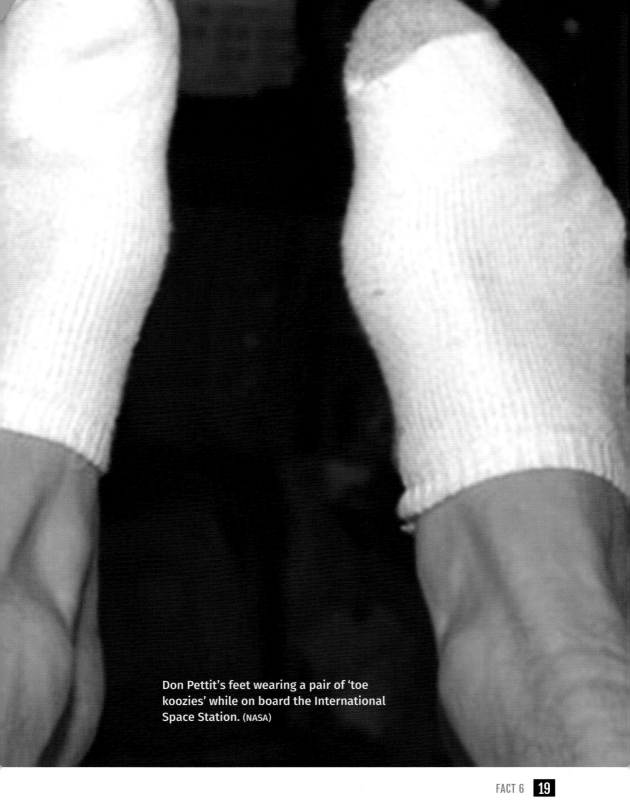

Don Pettit's feet wearing a pair of 'toe koozies' while on board the International Space Station. (NASA)

Ever watched a science fiction film where there is a massive battle scene? From Star Wars to Star Trek, the destruction of an enemy ship usually comes with a massive explosion sound. The truth however, is somewhat less dramatic and actually quite eerie. You may have heard the saying that in space, no-one can hear you scream! Indeed it is true that sound cannot travel through the vacuum of space so an exploding ship would come with no noise at all, not even a distant pop. That is, unless broken ship parts hit your ship, then you certainly would hear the banging as they crashed into you. To understand the silence of space, we need to understand how we hear sound.

The air that surrounds us at sea level contains, very approximately, 2.5×10^{25} molecules of gas (mixture of oxygen, nitrogen and a few others) per cubic metre. In contrast the space between the stars that is known as the interstellar medium has around 1×10^6 molecules per cubic metre in it. Sound should be thought of as the result of something moving and disturbing molecules around it, which is then detected by something sensitive to the movement of molecules. Consider someone talking: their vocal chords vibrate and cause a disturbance in air molecules. This disturbance propagates through the air as molecules knock into each other until eventually molecules near your eardrum get disturbed. Your ear drum detects the disturbance and you hear the sound.

The transfer of sound through disturbance of molecules works fantastically on the surface of the Earth because the molecules are plentiful, but out in space, molecules are much more scarce. Not many molecules means the disturbance does not get transferred and no 'sound' is heard. You could be floating next to someone in the depths of space and they could be shouting at you, but you would hear nothing. Even if you were somehow floating through a gigantic gas cloud in space, the birth place of stars, the average distribution of molecules means that you would still only find 1×10^{12} molecules per cubic metre, not dense enough for sound to travel through.

Whilst it is true that sounds cannot be heard in space this should not be confused in a discussion about radio waves travelling through space. Televisions and radios, for example, work by receiving radio waves from a transmitter which can be located tens or hundreds of kilometres away. They convert the information which is encoded in the radio waves into sound which pops out of a speaker. The radio waves that carry the information are an electromagnetic phenomenon that can travel through the vacuum of space just like visible light.

Astronaut Bruce McCandless floating above Earth in the eery quiet of space while the space shuttle *Challenger* was 320 feet away . (NASA)

FACT 8

Hot stars are blue, cold stars are red...our taps are wrong!

We tend to associate hot things with the colour red and cold things with the colour blue. After all, fire is associated with an orangey/red colour and ice seems to have a blue/white tinge to it. So convinced are we of this association that the tiny little caps on taps show red for the hot tap and blue for the cold. Alas, we are wrong and you only have to watch a piece of metal in a furnace or look at the stars to see it.

Gaining access to a furnace is somewhat tricky but suffice to say, leave a piece of metal inside one and it will first start to glow red and yellow and as the temperature increases turn to white and blue. This can be seen in the night sky too. Ask anyone what colour the stars are and most people will say with some conviction that they are white. Look up at the sky on a clear moonless night, carefully study them, and whilst the great majority will look white, a few of them will clearly look coloured: Betelgeuse in Orion is red, Capella in Auriga is yellow and Rigel in Leo shows a hint of blue.

The colour of a star, or an iron bar in a furnace is determined by its surface temperature, which is itself caused by energy exciting the atoms. As you may recall from your school days, atoms are made up of a nucleus of protons and neutrons (of varying quantities) and any number of electrons in orbit. The electrons can only exist in certain energy states around the nucleus and can be visualised as planets in orbit around a star. The orbits of the planets around a star are analogous to the orbit of electrons around the nucleus. Blast the electrons with some energy such as heat, and the electrons get excited.

Exciting an electron allows it to leap to a higher energy state, a bit like a planet moving to an orbit a little further away from its star. For the atom, this is an unstable state; the electron wants to drop back to where it came from and in doing so, it has to dump the excess energy. It does this by releasing the energy as photons of light. Stars (or pieces of metal) that have a higher surface temperature give more energy to the atoms, which excites the electrons even more. As the German physicist Max Planck showed, energy relates to frequency of light so that the higher the energy state of the electron, the higher the frequency of light it emits when returning to its stable state.

It seems then, that our perception that cold things are blue is quite wrong. This slightly misguided opinion probably stems from ice and water having a blue 'appearance' but that is largely due to reflection of the sky rather than anything to do with its temperature. Perhaps we should change our taps!

NGC 6397 is a fabulous star cluster composed of hundreds of thousands of stars; many are red or blue in colour. (NASA)

FACT 9

Some stars spin 700 times per second

We are all familiar with the concept of a day, it's 24 hours, right? Wrong, a day is nearly 24 hours it is exactly 23 hours, 56 minutes and 4 seconds. We just round it to 24 hours for convenience. The day is defined by how long our planet takes to spin once on its axis and we can see the tiny discrepancy in the stars. Look at the sky over a few nights at the same time every night and you will see that the stars slowly creep across the sky a little further every night. Not everything turns at this speed though, Jupiter for example takes 10 hours to rotate once but some objects spin considerably faster. Some neutron stars, the stellar corpse of massive stars have been clocked at spinning SEVEN HUNDRED times in one second! Before looking at why the stars spin so fast, it is worth briefly looking at what a neutron star is.

Stars do not last for ever, but they spend most of their lives converting one element into another deep in their core and for the most part there is a balance between the force of gravity trying to collapse the star and the thermonuclear force trying to expand it. For stars like our Sun, they produce carbon in their core and, because of their relatively low mass, eventually hit the end of the road and die. Stars at least eight times more massive can continue and create further elements in their core, even iron. A star with an iron rich core cannot sustain the fusion process and this ultimately leads to its gravitational collapse.

The collapse of a massive star is the cause of the insanely fast rotation. Think about an ice skater with their arms out, spinning on the spot. As the skater pulls their arms in their rate of spin increases in accordance with the conservation of angular momentum. Angular momentum defines the amount of rotation an object has and it is affected by the distribution of mass from the axis of rotation, or in other words, move the mass towards or away from the axis of rotation and the rate of spin will increase or decrease. It is easy to see why an ice skater spins faster as they draw their arms in because they shift the mass of their arms closer to the axis of rotation, so to conserve the angular momentum the rate of spin must increase; similarly move their arms out and they slow down. Try this for yourself on an office chair, being careful not to fly off of course. Stick your legs out as you start to spin then pull them in and you will go faster.

It is not just figure skaters or people on office chairs but everything in the Universe; any system must obey the conservation of angular momentum. As a massive star collapses at the end of its life, the great majority of material that made up the star gets compressed into a smaller object, and as it shrinks, the original rotation of the star must increase to conserve angular momentum.

3C58 is a supernova remnant that was observed in 1181AD. At its centre is a rapidly spinning neutron star. (NASA)

FACT 10

Venus and Uranus are upside down

Space has no up nor down yet by convention we class the 'top' of the Solar System as that within which we find the north pole of the Sun. If we observe the movement of the planets then we see that they all orbit around the Sun in the same direction as the Sun's rotation: we call this prograde motion. If they orbited in the opposite direction, like a number of comets, we would call it retrograde motion. We can also look at the planets and see in which direction they rotate and, for the most part the planets all rotate in the same prograde direction. Venus and Uranus, however, both rotate in the opposite direction.

To understand the retrograde rotation of Venus and Uranus we need to go back to the formation of the Solar System. The Sun and planets all formed out of a massive cloud of mostly molecular hydrogen 4.6 billion years ago. Forces would have been acting upon the cloud such as electrostatic forces between microscopic grains of dust or gravity from nearby stars, and as a result of these, the cloud would have had a little rotation. Over time, the cloud collapsed further and like a skater pulling in their arms the rotational speed increased, giving rise to the rotational speed of the planets we see today. This also explains why planets tend to rotate in the same direction that they orbit, except Venus and Uranus.

Exactly why these two planets rotate in the opposite direction is not known but a clue might be found if we look at the axis about which each planet rotates. Imagine the Sun sat on a massive sheet of paper; the eight planets would be orbiting around the Sun broadly upon this sheet. We call this plane the ecliptic. The amount of tilt of a planet's axis of rotation is measured in reference to the ecliptic. The Earth's axis is at 23.5 degrees to the vertical, Mercury is within a degree of being vertical but Venus is 177 degrees and Uranus 98 degrees. When we look at the two planets they look to be rotating backwards but in reality they are spinning in the same direction as the others, but they are just upside down.

There are numerous theories to explain why Venus and Uranus have been flipped over. It was once thought that they were struck by a large comet or asteroid in the early days of the Solar System. It is still thought this is the case for Uranus but Venus may have an alternate explanation. A once popular theory suggested that with its thick dense atmosphere and orbit around the Sun just 108 million km away, its atmosphere was subject to strong tidal forces. Atmospheric tides and the resultant friction could cause the planet to slowly flip over, giving it the 'upside down rotation' we see today.

Hubble Space Telescope image of
Uranus in visible light. (NASA)

FACT 11

Saturn is not the only planet with rings

One of the first things I saw through a telescope was Saturn. Its rings appeared beautifully to me through the 22 cm reflecting telescope. Since then I have seen all of the planets through a telescope, even elusive Mercury, but none were quite as remarkable as Saturn. If you are new to astronomy then you maybe surprised to learn that Saturn is not the only planet with rings. I have seen all of them telescopically but none display a stunning ring system like Saturn. The reason is simple. In comparison, the rings of the other giant planets, Jupiter, Uranus and Neptune, are feeble in comparison to mighty Saturn and require a space telescope or space probe to see them, but they are nonetheless fascinating to study.

The discovery of the rings of the other planets was suggested at by astronomers but not absolutely confirmed until space probes, chiefly the Voyager craft, visited them in the '70s and '80s. Even the discovery of Saturn's rings by Galileo using his (by today's standards) poor quality telescope was not clear cut. His first observations in 1610 reported that Saturn was a 'triple planet' and accompanied by moons which almost touched the planet. Over the following years he noted how the companions to Saturn seemed to change shape and even vanish from view. What he had not realised was that he had discovered the rings and their changing aspect to us here on Earth. Much like a slowing spinning top wobbles around on its axis, so does Saturn, thus presenting the rings to us at differing angles.

Despite the dominance of Saturn's rings, there are similarities among them all, mostly their composition. The rings of all the planets are made up of tiny particles that orbit around their parent planet much like the Moon orbits the Earth, yet their distance from us gives the illusion that they are solid structures. There are differences among them too: the rings of Jupiter, Uranus and Neptune are mostly composed of dust and chunks of rock while the rings of Saturn are made up of mostly ice and rock.

There is still much debate over the origin of the rings although they are clearly a common feature of massive gas planets. One theory suggests that the rings might have formed at the same time as the planets themselves. The presence of numerous moons surrounding the outer planets certainly shows there was excess material from the planets' formation and it may simply be that gravitational forces from the planet stopped the ring particles from forming into a moon and instead spread out along the orbit to form a ring. Perhaps a close encounter with another moon led to tidal interactions which broke the young moon into pieces that formed the ring. For now though, the origin of the rings of the outer planets shall remain a mystery.

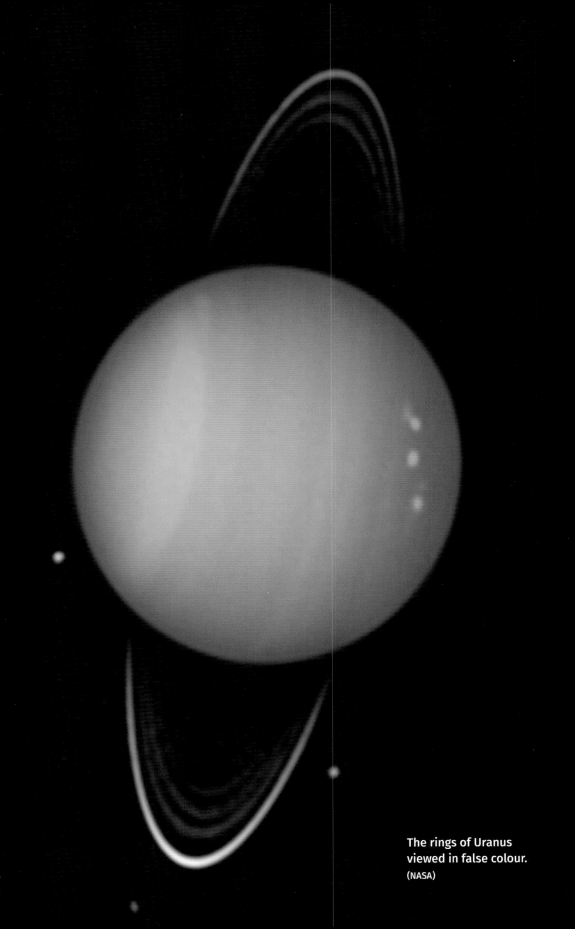

The rings of Uranus
viewed in false colour.
(NASA)

FACT 12

Dark sunspots are as bright as the full Moon

The Sun is a pretty average star fusing hydrogen to helium deep in its core. It appears so important and significant to us simply because it is close – relatively speaking. It is true that you should NEVER look directly at the Sun through a telescope or binoculars because the intense amounts of magnified energy can easily burn a hole in your eye. There are safe ways to study the Sun, however, and one of these is to project a magnified image of the Sun onto a piece of white card. If you do this (and you can try it yourself but check online for details of how to do it safely) then you can often see tiny dark blotches on the visible surface of the Sun. These sunspots look dark, even black, but in reality, if you could detach them from the rest of the Sun, they would shine brighter than the full Moon.

The Sun, like all other stars, is a massive ball of gas, or more accurately, ball of plasma. Plasma is known as the fourth state of matter and whilst it is similar in many ways to gas, it is subtly different. A gas is a substance that can expand freely to fill any space, whereas a plasma behaves in the same way but the atoms have had electrons stripped off them making them positively charged. The positive charge of plasma is key to the formation of sunspots.

The Sun, like the planets, rotates on an axis but since it is made of plasma, different regions of the Sun can rotate at different speeds. For example, the polar regions complete one revolution in 31 days whereas the equator completes a rotation in 27 days. This is known as differential rotation and it plays havoc with the Sun's magnetic field. Like our own magnetic field there are field lines that run between the north and south pole but on the Sun, the magnetic field runs through the plasma and as you recall this holds a charge of its own. Any movement of the plasma can drag the magnetic field with it, and over a period of a few weeks, the magnetic field can get wound up tighter and tighter by the differential rotation. As it winds up, field lines start snapping and bursting through the visible surface of the Sun and when they do they they inhibit the flow of energy from the Sun's interior, temporarily stopping it reaching the surface.

The visible surface of the Sun, known as the photosphere, has a temperature in the region of 5,500 degrees but these localised regions are cooler at around 4,000 degrees. In contrast with the hotter, brighter photosphere these sunspots appear visibly darker but in reality they are still emitting an incredible amount of light. If you could somehow detach the sunspots and somehow separate them from the brighter photosphere they would still be brighter than the full Moon.

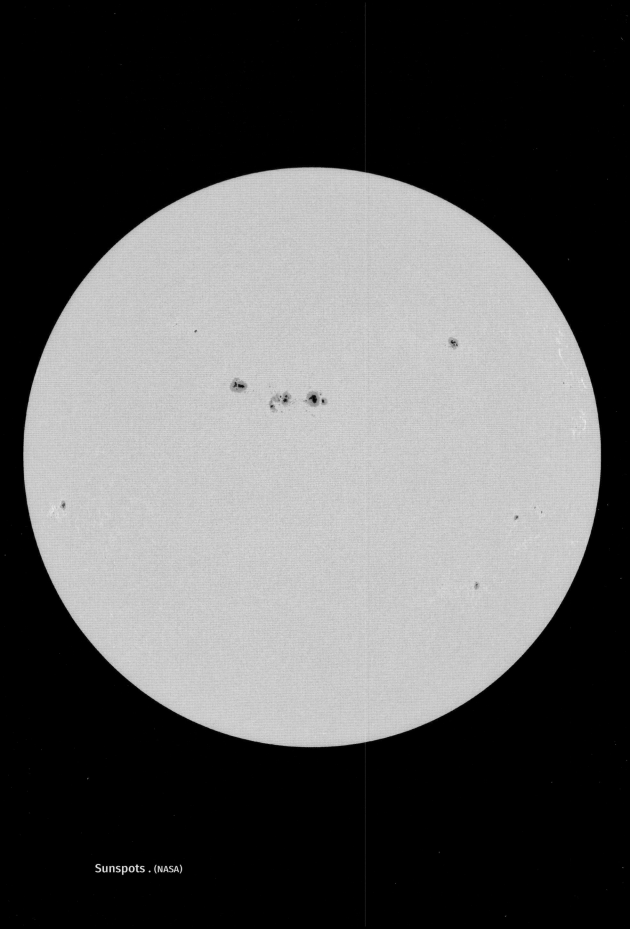

Sunspots . (NASA)

FACT 13

Metal sticks together in space.

It is a scene you may have watched before. An astronaut is out on a spacewalk, or extra-vehicular activity to be more precise, and they have tools strapped to their space suit while effecting some sort of repair or installation. In an activity like this there are lots of challenges, not least of which is keeping the astronaut alive, but at a more mundane level, using tools in space can prove to be a little challenging. If you have ever tried to adjust a nut with a spanner down here on Earth then you will know that generally, unless it gets stuck, you can use the spanner and then remove it from the nut. In space, nothing is that simple.

The atoms in a metal object share very strong bonds between them which keep them rigidly together. It might be reasonable to assume that placing two metal objects together might cause the atoms in the outer layers to instantly create new bonds with their new atomic neighbours, fusing them together. Instead we take it for granted that we can get the nut off the bolt or remove any metal item from any other metal item if they touch (unless it is a magnet of course) but there is science behind this. The atmosphere of the Earth has a high nitrogen content but crucially a lot of oxygen too. The oxygen in the atmosphere allows us to live and breathe but is also responsible for metal items not sticking together here on Earth in a process known as oxidisation.

Oxidisation is a chemical process you may have heard of but will almost certainly have experienced first hand. Ever left a metal object outside for a few days, weeks or even months and come back to it only to find it has rusted? Rusting is an oxidisation process where oxygen in the atmosphere has bonded with iron atoms in the metal by stealing a few of their electrons and producing iron oxide, or rust.

Oxidisation takes place when an oxygen atom bonds with another metallic atom in the surface of metal, creating a thin film on the metal's surface. It is that film which stops metal objects sticking together on Earth, but of course there is no oxygen in space so this process cannot take place. The result ... place two metal objects (with no oxidisation on their surfaces) against each other and the atoms in the metal will establish bonds, sticking them together firmly. It is for this reason that any metal tools used in space are covered by some form of non-metallic covering and spacecraft designers need to be very careful about using metal components close to each other. This problem was very real for engineers working on the Galileo space probe as it travelled towards Jupiter in 1991. They were unable to fully deploy the high gain antenna and this was put down to metal components having unexpectedly stuck together.

**The *Galileo* Spacecraft
before deployment
from the Space Shuttle
Atlantis. (NASA)**

FACT 14

Astronauts cannot burp in space

A journey into space looks like great fun. After all, you get to experience weightlessness which in itself must be worth the trip. Certainly there are things you can do in space that you cannot do down here on Earth but there are also things you can do on Earth that you just cannot do in space. Bodily functions are themselves really quite different in space: your feet go soft, you become a bit taller and simple tasks like drinking become complicated. There is one thing we all do on average three to six times after eating or drinking ... burping! From small barely noticeable burps to massive great belches they are a part of life and one we cannot avoid, unless, it seems, we are in space.

A burp is a perfectly natural bodily function and has a lot to do with when we consume food or drink. When we swallow, it is not just the food or drink that goes down into our stomachs, we also swallow some air with it. Our bodies are pretty clever things though and we do not need nor want air in our stomachs, that is what the lungs are for, so the body tries to get rid of the excess gas in the stomach by trying to push it back out again. The gas gets propelled out of the stomach and through the muscular tube known as the oesophagus and finally out of the mouth ... BUUURRRPP! You may have noticed how eating too fast or drinking fizzy drinks makes you burp more. That is because fizzy drinks contain carbon dioxide gas and eating too fast causes you to gulp down more air, and more air or gas means more burping.

Part of the process of burping requires the food and drink to separate from the gas in your stomach, and we can thank gravity for doing this for us. When you are floating around in space the experience is that gravity is not pulling on you (although it is but you are in a state of free-fall so cannot feel it in the same way) nor is it seperating the contents of your stomach. The result that the gas gets trapped inside your stomach and cannot escape, making you incapable of burping. If something causes gas to escape from your stomach in space then you have the rather horrible experience known as bommiting, a cross between a burp and a vomit as stomach content gets propelled out with the gas. It is not unusual for astronauts to propel themselves across a room of the space station by pushing against it with their feet. This acceleration simulates gravity, separating out gas from content and allowing a rather more natural burp to come out.

Astronauts eating a meal on board the International Space Station. (NASA)

The days really are getting longer

I don't know about you, but for me, the days seem to get shorter and shorter. Seemingly less time to perform the seemingly constantly growing 'to do' list. Ask anyone and they will tell you that a day is 24 hours, and indeed it is the ancient Egyptians we can thank for this. They are responsible for dividing a day up into 24 hours: 10 hours of daylight, an hour of twilight either side and 12 hours of darkness, although interestingly the length of an hour varied throughout the year. If we accurately measure the rotation of the Earth against the stars then the same star comes back to exactly the same point in the sky 23 hours, 56 minutes and 4 seconds later. This has not always been the case: throughout much of the Earth's history the rotation of the Earth has been slowing at around 25 millionths of a second every year. The rotation of the Earth really is slowing down and the days really are getting longer. We can look to the Moon and the force of gravity for the explanation.

Any object in the Universe which has mass has an associated force of gravity that we perceive as a pull on objects around it. It is gravity that 'pulls' on us to keep us pinned to the surface of the Earth. The same gravity from Earth pulls on the Moon and keeps it in its orbit, but at the same time, the Moon's gravity pulls at Earth. The effect of this pulling at Earth is a tidal bulge that should sit directly between the Earth and Moon. The bulge occurs across land and sea and has varying amplitudes dependent on many different factors such as depth of sea, local geography and even the position of the Sun relative to the Earth and Moon. In reality, the tidal bulge is not directly between the Moon and Earth. As the Moon slowly orbits the Earth, the Earth spins faster underneath it. The faster rotation of the Earth tends to drag the tidal bulge a little ahead of the Earth–Moon line and it is this which has a rather dramatic, yet slow effect.

Fundamentally, the tidal bulge is a deformation of the shape of the Earth but in reality, it is a whacking great lump of mass which, like all mass has gravity. The gravity coming from the tidal bulge pulls on the Moon and causes it to accelerate a little in its orbit. The acceleration of the Moon in its orbit causes it to slowly move further away at a rate of four centimetres per year. At the same time, the gravity of the Moon pulls on the bulge of Earth, slowing the Earth's rotation by a tiny amount. The effect is small though and it will be another 140 million years before a day is 25 hours long!

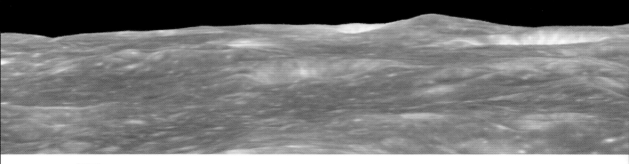

FACT 16

Neptune is home to the fastest winds in the Solar System

Wind is a familiar meteorological phenomenon here on Earth. No matter where you live, you can be sure that at some point you will feel a refreshing breeze or warm tropical air wafting by you. The fastest wind gust speed recorded on Earth (that was not part of a hurricane or similar event) was 408 kilometres per hour by a weather station in Barrow Island in Australia in April 1996. Compared to some of the winds clocked up on other planets though, this is nothing more than a breeze.

As long as the energy from the Sun is beating down on us we will always have weather of some sort and will always have wind. Energy from the Sun travels the 150 million kilometres to us here on Earth and on arrival travels through the atmosphere. Eventually it hits the surface, be that land or sea, and slowly starts to warm it up. As the surface warms, it starts to reradiate energy out again and it is this which finally warms the atmosphere. Any parcel of air in contact with the surface gets warmed up and as it does, the molecules in the air gain energy and start moving around faster. With their increased speed, the parcel of air expands and its density decreases causing it to be less dense than surrounding air. It starts to rise and as it does, more parcels of air drift across the surface to fill the void left by it. This is what we experience as wind. The exact conditions in the atmosphere will determine exactly how fast the air moves across the surface.

It is easy to measure wind speed on Earth because we can install anemometers to directly measure the flow of air but on other planets we cannot do that. Instead, measurements of cloud movement can tell how fast wind is blowing, and on Neptune, clouds have been measured moving at speeds in excess of 2,200 kilometres per hour. The nearly supersonic wind speeds on Neptune have perplexed scientists for quite some time but recent studies revealed that the winds seem to be constrained to thin layers of the upper atmosphere. More detailed analysis shows this region to occupy the outermost 0.1% – 0.2% of the atmospheric layers, which leads to the conclusion that the driving force behind the winds is condensation and evaporation of moisture due to solar heating. For now, this is just conjecture and more research is needed to explain the extreme wind speeds in Neptune's atmosphere.

The clouds of Neptune revealed in amazing detail by the *Voyager 2* spacecraft. (NASA)

FACT 17

There is no dark side to the Moon

The concept of the Moon having a 'dark side' is not a new one; many children grow up thinking there is a dark side to the Moon, and indeed the concept has found its way into films and even music. In reality, the Moon has no dark side. It has no 'bright side' for that matter but what it does have is a 'far' side. It seems strange to talk of the Moon having any sort of 'side' because it is a sphere; perhaps it is more accurate to say it has a far hemisphere, meaning one portion of the Moon is not visible from Earth, at any time.

You are probably already familiar with the phases of the Moon and have more than likely seen the full Moon, thin crescents or the so called gibbous shaped phases too when over 'half' of the Moon can be seen in the sky. If you have studied the motion of the Moon across the sky and the times when the various phases are visible then you may have come to the conclusion that the phases are linked to the relative positions of the Sun, Earth and Moon in the sky. We see full Moon when the Earth-facing hemisphere is fully illuminated and the other hemisphere would be in darkness. Conversely, a new Moon would appear dark to us and the 'far' hemisphere would be illuminated by the Sun. During the Moon's journey around the Earth both hemispheres get illuminated by the Sun, but one hemisphere of the Moon always faces Earth.

In an earlier fact we saw that the gravitational pull of the Moon causes a tidal bulge to appear on Earth and that this lies slightly ahead of the line between the Earth and Moon. It is not just the gravity of the Moon causing tides to appear on Earth, but the Moon too experiences tides from the gravity of Earth. In the distant past, the tidal bulges created on the Moon tended to lead ahead of the line between the Earth and Moon. The gravity of Earth would tug against this bulge and, over millions of years, would slow the rotation of the Moon until one hemisphere of the Moon was tidally locked to Earth. Once this happened, the Moon's axial rotation would match its orbital period and it would forever complete one orbit of Earth in the same time it takes to rotate once on its axis.

This tidal locking is not unique to the Earth and Moon system. There are around 200 moons in our Solar System and many are tidally locked to the object they orbit. If it were not for the influence of other objects in the Solar System, the slowing down of the Earth's rotation due to the pull of the Moon would ultimately lead to the Earth becoming tidally locked with the Moon, but this would not happen for another 50 billion years or so!

In practice, we do see a little more than just 50% of the surface of the Moon but this is due to the elliptical and tilted nature of the Moon's orbit around Earth. For example, the orbit of the Moon is tilted by five degrees to the orbit of the Earth around the Sun so occasionally the Moon is 'lower' than usual, allowing us to see a little further over the north pole of the Moon, and on occasion it is a little higher, allowing us to see further beyond the south pole. Taking into account the various effects of the Moon's orbit, we can see 59% of the surface of the Moon (not all at once) but there is 41% of the Moon's surface that can only ever be explored from space.

The far side of the Moon is revealed
by this image by NASAs Lunar
Reconnaissance Orbiter. (NASA)

FACT 18

Black holes are not actually holes!

Look up at the sky on any night and depending on the amount of lighting near you and on the phase of the Moon you will be able to see anything from just a few to almost 2,500 stars. At first glance they will look pretty similar except for a few that look brighter than the others they vary in colour, but by and large, there's not a lot of variation in their appearance. In reality stars vary by an enormous amount, from young hot stars, to old cool stars, stars which vary in the amount of light they give off and even stars that are dying, or are dead. It is these dead stars where we find the exotic type of object known as a black hole.

Their name suggests they are some form of cosmic hole but the nature of these cosmic corpses is far more interesting. Fact 3 showed how massive stars synthesise heavy elements deep in their core through the process of nuclear fusion. The most massive stars will end up with a core of iron atoms but it is not possible to fuse iron in the core of stars. With no fusion in the core, the resultant drop in the thermonuclear force allows gravity to win and the core of the star collapses. For the most massive stars, the core of the dead star is subject to such immense crushing pressure that it is compressed down to an object of no dimensions: a singularity. Singularities are infinitely dense and it is their effect on the surrounding region of space that gives them their name.

Imagine for a moment you are holding a ball and throw it up into the air. Naturally, gravity stops the ball from flying off into space, but if you could throw the ball at 11 km per second it would escape out into the Solar System. The speed at which the ball needs to travel is determined by the strength of the pull of gravity so if you were to try the same exercise on Jupiter you would have to throw the ball at 59.5 km per second. Try the same thing on a black hole and you would have to throw the ball faster than the speed of light, in other words faster than 300,000 km per second.

If you were, by some miraculous quirk of the laws of physics, able to stand on a singularity and turn your science busting torch on, then the light from the torch would only get so far before gravity would stop it escaping. This point is known as the event horizon and it marks the boundary of the black hole. Anything that goes on within the event horizon or falls beyond it, can never escape nor can it be observed by the outside world. It is not surprising that these objects have earned the name 'black holes' given that no light can escape from them and anything that falls in can never get out!

The multitude of stars in the direction of the centre of the Milky Way where a massive black hole is thought to reside. (NASA)

FACT 19

Clusters of galaxies are used as gigantic cosmic telescopes

Think of an astronomer and it is not long before the trademark telescope also pops up in your head. Telescopes usually have either a lens or mirror or combination of both which focusses light from distant objects into an eyepiece for observation, or onto an instrument of some sort for further study. Even if you do not have access to a telescope you have still probably seen through one at some point in your life, even if it is only one at a seaside resort for looking out to sea.

There are many different types of telescope for studying different types of objects: infra-red telescopes, ultraviolet, radio and even gamma ray instruments, and all of them use something to bend or focus incoming light from distant objects. Albert Einstein was the first to predict that light could even be bent by gravity! In 1919, Sir Arthur Eddington led a trip to photograph a total eclipse of the Sun, during which the bright photosphere of the Sun gets obscured by the moon. With the darkened Sun it was possible to capture photographs of stars whose light had passed close by the eclipsed Sun and as predicted by Einstein, their positions had shifted by a tiny amount. Einstein was right, gravity did bend light.

In 1979, two quasars (a massive and extremely distant astronomical object that emits exceptional amounts of energy) were discovered very close to each other. Further study revealed that they were at a similar distance from us and had almost identical spectra. The two quasars turned out to be two images of the same object. These were the first gravitationally lensed observations. A gravitational lens works just like a conventional lens but instead of being made from glass, the 'lensing' item is generally a massive galaxy cluster.

The mechanics of a gravitational lens are probably easier to understand if you imagine space as a giant sheet of rubber. Place a bowling ball on the rubber and you can visualise how the rubber will bend under the mass of the bowling ball. Any smaller balls being rolled along the rubber sheet will, when they get close to the bowling ball, follow the distorted curvature of the rubber and their path will be bent. In just the same way we can think of a galaxy cluster distorting the fabric of space, such that any light that tries to pass close by will get bent by it, with the amount of bending determined not just by the mass of the intervening galaxy cluster but also by the distribution of that matter and the relative position to the line of sight of the intervening cluster. Some incredible shapes can be formed by these gravitational lenses from complete circles to crosses and arcs, all of which will be distorted images of the distant galaxies that have been 'lensed'.

Abell 2218 is a galaxy cluster 2.1 billion
light years from Earth and a great
example of a gravitational lens. The thin
curved arcs are lensed images of more
distant galaxies. (NASA)

FACT 20

Mercury is the fastest planet

It is not something that is often thought about but Earth and all its inhabitants are hurtling through space at an average speed of 30 kilometres per second. Different planets orbit the Sun at different speeds: for example outermost, Neptune, travels at 5.4 kilometres per second, Jupiter 13 kilometres per second and Mars at 24 kilometres per second. As you may have noticed there is a correlation between a planet's distance from the Sun and the speed it travels around the Sun, so clearly planets that are closer to the Sun than Earth are going to be travelling even faster. Venus, the second planet from the sun, travels at 35 kilometres per second but the record holder is Mercury, which rattles around the Sun at the breakneck speed of 47 kilometres per second.

In fact number six we already saw how a ball dropped from your hand falls to Earth under the force of gravity. At the distance of your hand from the centre of the Earth the pull of gravity exerts a certain pull on the ball. Move your hand higher (or further away from the centre of the Earth) and the pull of gravity is weaker, move it closer and the pull is stronger. It is this differing pull of gravity which determines the speed of a planet in its orbit.

Fact six went on to explain that if you throw the ball it would follow a curved path to the ground, throw it harder and therefore with greater speed and the curved path will be shallower but still it will follow a curved path before landing. It is easy to see that there must be a speed you could throw the ball so that it follows a curved path towards the surface of the Earth but as it falls, the surface of the Earth curves away from the falling ball. The ball is then in orbit. If the same experiment were performed further away from the centre of Earth then the ball would be subject to less pull from gravity and would not have to be thrown quite so hard for it to remain in orbit.

The planets are all travelling around the Sun and in orbit and they, like the ball, are in this state of 'free fall' around the Sun, constantly falling towards the surface of the Sun but with the curve of the Sun falling away from them. The closer the planet, the faster they must go to remain in orbit. It is a little more complex than this in reality because the orbits of the planets are not perfectly circular, they are elliptical, more like a slightly squashed circle. When they are closer to the Sun they move faster and when they are further away they move slower, so the values quoted earlier for their orbital speeds are averages.

Mercury, the nearest planet
to the Sun imaged by the
Messenger spacecraft. (NASA)

FACT 21

On Mars you need flip flops and a hat

Ever wondered why it gets cold the higher you go? If you have climbed a mountain then you will be familiar with it being a bit chilly at the top but nice and warm nearer the ground. If you have been on an aeroplane then you may not be aware but at the altitude most airliners operate at, around 40,000 feet (about 12 km), the temperature plummets to on average -50 degrees! The decrease in temperature with height is something many people probably just accept but if you went to Mars, the decrease in temperature with height is nothing short of extreme.

Before looking at the conditions on Mars let us first look at the cause of general atmospheric temperature change. The atmosphere on Earth, just like the atmosphere on Mars, is a mixture of gasses. On Earth it is mostly nitrogen but with significant amounts of oxygen and other gasses. On Mars it is almost entirely carbon dioxide, a little nitrogen and even smaller amounts of other gasses. In the case of Earth, the atmosphere is approximately 480 kilometres thick so atmospheric gasses at ground level have 480 kilometres of gas pushing down on them. You can experience something similar by diving to the bottom of a deep swimming pool and feel the pressure from the water above you. Go a little higher and there is less gas above and the result is that an increase in height

An image of Bonneville Crater on Mars, taken by *Martian Spirit* rover. (NASA)

brings with it a reduction in atmospheric pressure. A decrease in the pressure of a gas brings with it lower temperatures. This may be something you have noticed if you hold your finger over the end of a bicycle tyre pump when you pump it. By blocking the hole the air inside cannot escape, the pressure builds and you feel the pump nozzle get hot. Similarly emptying a pressurised can, which decreases the pressure inside, leads to the can feeling cold. It is this mechanism, decreasing pressure, which causes temperature to reduce with height.

A thick planetary atmosphere like the one here on Earth can also have an insulating effect so that the heat from the Sun gets trapped and does not easily get lost to space. The Martian atmosphere is about 100 times thinner than Earth's so any heat the Martian surface receives from the Sun soon gets lost. This can be seen in the extreme temperature differences during the summer day and night on Mars where at the equator, the temperatures can reach a pleasant 20 degrees in sunlight or a rather more nippy -73 degrees at night. If you were able to stand on the surface of Mars without the protection of a spacesuit then you would find that you have nice toasty feet but your head would be cold enough to need a hat.

FACT 22

The moons of Jupiter told us the speed of light

When Galileo turned his telescope on Jupiter over 400 years ago, he discovered that the mighty planet was not alone: it had four tiny moons in orbit around it. These moons became known as the Galilean Satellites and are called Io, Europa, Ganymede and Callisto. The discovery of the moons of Jupiter became one of the findings that proved the Earth did not sit at the centre of the Solar System. Careful observation of the moons also revealed something really quite unexpected and led to the first calculation of the speed of light.

The discovery came in 1676 by the Danish astronomer Ole Romer, but prior to his observations, others had tried to calculate the speed of light through experiments on Earth. Galileo himself had tried to determine the speed of light using shielded lanterns some distance away. His assistant would uncover the lanterns at certain times so that Galileo could measure the time it would take for the light to reach him. After a number of unsuccessful attempts he concluded that light travelled at a speed too fast to be determined by the experiment. It later transpired that the lanterns were just under 2 kilometres away so the light would have taken 5 millionths of a second to travel the distance between lantern and Galileo, a period of time he could not hope to measure with the equipment of the day.

Jupiter and two of its
satellites Io (left) and Europa
(right) taken by the *Voyager 1*
spacecraft. (NASA)

Romer on the other hand had more success by observing the moons of Jupiter, in particular the nearest of the Galilean moons, Io. Io orbits Jupiter once every 42.5 hours and the plane of its orbit around Jupiter is very close to the plane of Jupiter's orbit around the Sun. This means that Io is often eclipsed by Jupiter and it was these events that Romer observed.

Romer was timing the interval between successive eclipse events and noticed something quite remarkable. The eclipses, or rather the reappearance of Io from behind Jupiter should occur at regular intervals but sometimes they seemed to be a little later than expected, at other times a little earlier. Taking the assumption that Io's orbit was regular, Romer correctly theorised that it was as a result of the changing distance between Earth and Jupiter. When the distance between Earth and Jupiter was at its greatest, it took extra time for light to travel the further distance and when the distance was less, the the timings were reduced.

The orbits of Jupiter and Earth were well known, so, by taking the time difference into account, Romer was able to calculate an approximation for the speed of light. He came up with the value of 225,000 kilometres per second, which was very close to today's accepted value of 299,792 kilometres per second.

FACT 23

Earth is travelling through space at 225km per second

It is very easy to understand why our predecessors developed the perception that Earth was at the centre of the Universe and that all other astronomical bodies revolved around us. Even casual observation of the Sun, Moon, stars and planets reveals them following a path across the sky that appears to encircle our home planet while it, perceptibly, stands still. Through careful observation, our earliest astronomers started to notice things which put doubt in their minds that the Earth was at the centre. The changing phases of the Moon, the retrograde motion of Mars across the sky and the moons in orbit around Jupiter are just some of the observations that made them question their theories.

The concept of the Earth moving through space and not sitting at the centre of the Universe is hard to accept but this latter misconception was disproved by a number of astronomers over the years, but of note Harlow Shapley who studied globular clusters in the early 1900s. By measuring the brightness of stars in the clusters or by measuring the brightness of clusters themselves when individual stars were not visible led him to the conclusion that globular clusters were arranged in a massive halo that centred around a point 3,260 light years away. This point marked the centre of the Milky Way galaxy and although the currently accepted figure is considerably more at 26,490 light years, Shapley's crude approach still proved that our Solar System was quite some distance from galactic centre.

It is a simple assumption to make that if our entire Solar System were 26,490 light years from the centre of the Galaxy then it must also be orbiting the centre of the Galaxy. The challenge comes in proving and measuring such motion. It is easy to measure the speed of a car for example because the driveshaft not only turns the wheels but also turns the speedometer cable which measures speed.

Thankfully the Universe has its own concept of a speedometer cable and we can find it buried in light. If you pass the light from a nearby star through a spectroscope you can see dark lines superimposed against it. These dark lines are there because of the presence of certain gasses, but if the object is moving, the lines will shift. Measuring the shift (known as doppler shift) allows us to calculate the speed of the object we are studying, or more accurately, the speed of that object relative to us. British astronomer William Huggins was the first to measure the velocity of a star using this doppler technique, and studying the motion of stars around the night sky allows us to conclude that Earth and the Solar System are hurtling around the centre of the Galaxy at 225 kilometres per second.

Studies of globular clusters such as M23 reveal our distance from the centre of the Galaxy and hence the speed that we are travelling. (NASA)

FACT 24

The light from stars tell astronomers what the Universe is made of

If you want to learn about the nature of an object on Earth then you can walk over to it, pick it up, analyse it and even subject it to complex experiments to probe its inner secrets. For astronomers, it is not quite as easy as that, because the objects under study are so far away it takes many years, even millions or billions of years for the light to reach us. Instead, astronomers have devised cunning ways to unlock the mysteries of the Universe by careful study and observation.

One of the ways astronomers can do this is to look very carefully at the light being received from astronomical objects, because deep inside hides a secret code just waiting to be unlocked. To unlock it astronomers need to use an instrument called a spectroscope which separates the incoming light into its component parts known as a spectrum. You will have seen a spectrum before, when our sky is graced by a rainbow, which is a very simple spectrum of the Sun.

Spectroscopes allow us to look at the spectrum of the Sun, or stars and even galaxies in much more detail, and when William Hyde Wollaston observed the solar spectrum in 1802 he saw dark lines superimposed on the colours. He had not quite appreciated the importance of his discovery which came some years later when Joseph von Fraunhofer continued studies of the solar spectrum. Fraunhofer also discovered absorption lines in stars leading him to the conclusion that they originated in the Sun and stars and carried valuable information about the nature of the object.

The absorption lines arise as a result of changes in the atomic structure of atoms. The general structure of an atom is a nucleus composed of protons and neutrons, with electrons orbiting around them. For example helium gas atoms have two protons and two neutrons in their nucleus and will, if they are not charged, have two electrons in orbit. To visualise the orbits of electrons, imagine the planets in orbit around the Sun: Earth is in one orbit and Mars is in another orbit a little further out. In the case of the planets, they tend to stay in their orbits, but electrons can hop between orbits around the nucleus. To do so they either need energy to go to a higher energy state (further orbit) or release energy to return to their natural energy state (normal orbit). Light, or more accurately electromagnetic energy can provide electrons with energy to boost them into a 'higher energy state' and this energy transfer manifests itself in absorption lines. Just like the barcode on an item of shopping which has a specific set of lines, different types of gas atoms have specific sets of lines too, so by studying what lines can be seen, we can tell what gasses are present.

The solar spectrum revealing absorption line details. (NASA)

FACT 25

Space agencies use planets to steer planets around the Solar System

When Galileo, Newton and Einstein came up with their laws governing how gravity manifests itself little did they realise that their work would aid the exploration of interplanetary space. From the early days of space exploration with the launch of the Sputnik satellite to the later and more complex Voyager interplanetary probes, a fine understanding of gravity has been essential to the success of the missions. Even the launch requires an understanding that a speed of at least 11.2 kilometres per second is needed to escape the gravitational pull of Earth. Escaping from the gravity of Earth is only part of the challenge, getting to the outer planets requires a massive amount of energy to travel against the gravity of the Sun and is a whole different story.

At the distance of the Earth–Moon system, the escape velocity from the Sun is 42.1 kilometres per second, so to get a spacecraft from Earth to the outer planets requires lots of energy. One option is to carry enough fuel to propel the rocket to suitable speeds but if you carry more fuel then you must launch the extra fuel off the Earth which requires even more fuel. It is a very inefficient way to travel, so instead, we can use the energy stored up in the planets' motion around the Sun in what is known as a gravity assist manoeuvre.

The concept of gravity assist is simple to understand. Imagine you are standing with a tennis racket and a ball is heading towards you. In the simplest model, when you strike the ball, it can head back in the opposite direction faster than when it arrived. The ball stole a little energy from your tennis racket. The result, the ball went quicker and the racket slowed down a little bit. The racket is a lot bigger than the ball so it only lost a small amount of speed to give the ball a lot of extra speed. Gravity assist is similar.

A space probe destined for Saturn would be travelling at a certain speed following launch. It would not have enough speed to get out to Saturn so its first destination is a fly by of Venus, a planet closer to the Sun. As it flies by Venus it gets accelerated to a higher speed but the laws of the conservation of momentum dictate that for the space probe to speed up, Venus must slow down. The enormous difference in mass between Venus and the probe mean that Venus slows down an almost imperceptible amount in the speed of its orbit around the Sun to give the space probe a decent speed increase. Using the gravity of planets can not only accelerate probes to higher speeds to get to the outer Solar System but can also be used to adjust the trajectory of the probe. In the case of the Voyager spacecraft, they had a very specific launch window that enabled them to fly by the planets on the way out of the Solar System so that each planet directed them on to the next.

The Voyager spacecraft. (NASA)

FACT 26

Shepherds keep the rings of Saturn in order

It has been over 400 years since Galileo first pointed his telescope at the planet Saturn. The crude nature of the optics meant that his view was less than perfect and he noted that it appeared to be a planet composed of three objects, one large and two smaller either side. Over the months and years that followed further observation revealed that on occasion, the companions to Saturn seemed to vanish from view. It was another 45 years after his first observations that Christiaan Huygens suggested the planet was surrounded by a giant disk. Galileo did not realise it but he had discovered the rings of Saturn.

Close up study of the rings has revealed that far from a disk surrounding the planet, the rings are millions upon millions of tiny chunks of ice and rock orbiting the planet just like the Moon orbits the Earth. From the distance of our vantage point they look like a solid, thin, disk except for a few gaps which are visible through telescopic observation. Even an amateur telescope will reveal gaps in Saturn's rings, such as the Cassini Division which separates the A and B rings, and the Encke Gap inside the A ring itself. When Galileo noted that the companion objects had vanished from view, he had witnessed a ring plane crossing. These events occur every 15 years when the Earth passes through the plane of the rings, causing them to momentarily be very difficult to observe and almost vanish from view.

The gaps and divisions in the rings of Saturn are the result of interactions with some of the many moons in orbit around the planet. The Cassini Division is the most prominent and is caused by Saturn's moon called Mimas. This moon (which rather looks like the Death Star from Star Wars due to the large crater which is a third of its diameter) orbits Saturn once every 23 hours but for every orbit that it completes, particles on the inner edge of the ring complete two orbits. The result of this orbital resonance is that the constant tugging from the same direction moves the particles out of the gap keeping it clear. Similarly, particles near the outer edge of the ring are in an orbital resonance where they complete three orbits for every one orbit of Mimas and again, they get cleared out of the ring through regular tugs from the moon.

Moons like Mimas are called shepherd moons and get their name from their 'herding' of ring particles keeping the rings in shape. Saturn has several shepherd moons, including Prometheus, Daphnis, Pam, Janus and Epimetheus. Any particles that start to wander out in front of a shepherd moon get slowed down by its gravity causing them to fall back in towards Saturn, while those that drift out behind a moon will get accelerated and thrown outwards away from the planet and into the outer rings.

The rings of Saturn clearly showing the Cassini Division between the bright B ring and dimmer A ring. (NASA)

FACT 27

The universe is expanding faster now than in the past

Learning how the Universe has evolved is easier than you might think. Imagine standing in a forest looking at all the trees. As you glance around, you can see saplings, strong mature trees and even trees that have died and fallen to the forest floor. You do not have to stand there for a long period of time to watch the lifecycle of a tree to quickly deduce how they live and die. In the same way, we can look at the Universe around us today, at the multitude of objects we can see in the sky and deduce how the objects in the Universe evolve.

Stars are a great example and to understand their lives we can study their nurseries in great clouds of nebulosity, can study stars in their adulthood and can even study dead stars. It takes light millions and sometimes billions of years to traverse the great distances in space so when we study distant objects we look back in time and can study the early Universe too.

Such studies have allowed us to understand how the Universe has evolved and even to calculate how fast it has been expanding. Until 1998 it was thought that the expansion of the Universe might eventually cease and ultimately collapse, or continue to expand forever, slowing down gradually but never stopping. That all changed in 1998 when astronomers were studying distant supernovae explosions. These stellar events mark the death of supermassive stars and are very useful in calculating the scale of the Universe.

There are a number of different types of supernova explosion and each one outputs a certain amount of energy. If we observe a supernova in a distant galaxy, we can determine which type it is by studying the way its light changes over time and can therefore deduce how much energy it should be expelling. Comparing its actual output from its observed output allows us to determine the distance to it. When astronomers did that back in 1998 they found something quite remarkable. They found that the supernovae were further away than they expected them to be. The only conclusion was that the Universe was not only expanding but was expanding at an accelerated rate.

The nature of the accelerated expansion is still open to a lot of debate. Clearly there is some force that is driving the expansion and it has been given the name 'dark energy'. For now though, we have no idea what it is. One theory suggests virtual particles are popping in and out of existence and each time, exerting a tiny pushing force on the very fabric of space. If these virtual particles are popping in and out across the entire Universe then it may be that enough of a force is being generated to accelerate the general expansion of space. It does mean, that the Universe is now not likely to end in a big crunch, nor generally slow down to a crawl but instead is likely to ultimately rip itself to pieces! Don't worry though, if this is the fate of the Universe, then we have billions of years to go before that is likely.

Supernova remnant G299.2-2.9 taken by the
Chandra **X-ray observatory. (NASA)**

FACT 28

Zero gravity makes copulation rather tricky

Isaac Newton of gravity fame, came up with three laws of motion. The first states that an object will remain at rest or in motion in a straight line unless acted upon by another force. The second law states that the acceleration of an object (change in velocity or change in direction) is dependent on the force acting upon it and its mass. The third law states that every action has an equal and opposite reaction and it is this law which poses fundamental problems for the future of humanity.

Blaming our future problems on laws discovered by Newton may seem like a sweeping statement but in about five billion years the Sun is likely to swallow up the Earth, and if humans are to survive then we need to conquer space. Taking our species out into space holds many challenges like simulating the effect of gravity but if that is not easily achievable then copulation is just one of the activities made more difficult, and it is this where Newton's third law puts a spanner in the works.

Newton's third law, as we saw earlier, states that every action has an equal and opposite reaction. In the throws of copulation a simple thrust could send your partner across the room and yourself off in the opposite direction. Unless you are clinging on to each other in a vice-like grip then before very long you will be floating gracefully apart in the weightlessness of space and that is not helpful as you try to do your bit to 'ensure the survival of the species'. It is something we take for granted, but in space, while floating around, trying to get intimate with someone is just tricky.

Thankfully NASA and other space agencies are investing money into making the act of love an easier one while travelling through the weightlessness of space. The most successful solution so far is the so called 2Suit, which was designed by Vanna Bonta in 2006 and rather resembles an elaborate onesie. In essence, two lovers would both wear the 2Suit, which has strategic openings and velcro fasteners that can secure the wearers to each other. It was tested on a parabolic aircraft flight which undertook successive steep climbs and dives high in the atmosphere such that each dive brought about a short period of weightlessness for its occupants. The test took place in 2008 by Vanna and showed it to be cumbersome but moderately successful.

The suits, if further developed will not only be able to provide for some much needed intimacy on long space flights but can also be used to hold groups of people together engaged in group tasks or where another's assistance is required. They can also be used to tether individuals to resting stations during downtime. There is no doubt that our future is among the stars, but whether the 2Suit will be the an essential item for lovers to pack remains to be seen.

The sight of a larger than usual Moon rising is just an illusion

We see the Moon because it reflects light from the Sun and we see its changing phases because the angle between the Earth, Moon and Sun changes. Because of this, the full Moon always rises when the Sun sets, but have you noticed how the full Moon always looks massive when it's low down on the horizon? The more you see it the more you can convince yourself that it really is bigger than usual and mysteriously shrinks as it gets higher in the sky. It might surprise you to learn though that it is just an illusion and is no larger when it is lower down than when higher in the sky.

The illusion has become known as the Ponzo Illusion after psychologist Mario Ponzo who explained it in 1911. To understand the illusion think about the view that greets you as you look out across a mountain range. Mountains in the distance will tend to look a little more grey than their nearer counterparts because of atmospheric effects and our brain takes that cue and tells us they are more distant. In other words, our brains work out the distance to objects based on visual cues. It is not just the colour of objects that we use to infer distance but as Ponzo explains, we subconsciously work out the distance to objects based on their background or the setting in which we see them.

In the case of the Moon and its appearance low on the horizon compared to how it looks higher in the sky, the difference is what else we see it next to. Our brains have been conditioned to see things on the horizon as being further away so when we see the Moon low down near the horizon it is also close to 'those things further away'. Our brain therefore concludes that the Moon too must be further away than usual. This is where the illusion bit kicks in. We incorrectly deduce that the Moon is further away when it is near the horizon yet the size of the Moon's image on our retina is the same as always. Our brain therefore leaps to the similarly incorrect assumption that, because it the Moon is further away than usual yet producing the same size image on our retina as usual, then it MUST be larger than usual. If it was not larger than usual then it would appear smaller in the sky because it is further away. Once the Moon drifts higher in the sky, we have nothing to compare it with and the illusion slowly fades away with the Moon 'returning' to its normal size.

If you are not convinced of all this then you can try this simple test. Look at the Moon when it is near the horizon but find some way to look at it upside down. Doing this means the objects on the horizon no longer look familiar to you, and you look at them and the Moon more objectively allowing you to see them as they should be seen.

Many people will have grown up with the nursery rhyme 'Twinkle twinkle little star, how I wonder what you are...' The image it conjures of stars is that they sit up in the night sky sparkling away '...like a diamond in the sky'. The appearance of stars is perhaps not so different to the nursery rhyme, since they do indeed appear to twinkle, often so much that they can appear to flash different colours. The origin of the twinkling or astronomical scintillation to use the scientific term has a rather more terrestrial origin.

Typically, stars do not change their output of light quite as fast as the twinkling we can see. There are stars however, whose output of light varies. They are known as variable stars but their output generally varies over a few hours/days/weeks or years. A particular type of star known as a pulsar can change its output of light much more rapidly, even thousands of times per second, but these are not visible without optical assistance. Instead, the flickering of stars in the sky is caused not by the star itself but by our atmosphere.

Earth is surrounded by an atmosphere which is a shell of gasses on average 100 kilometres thick, and any light from astronomical objects must travel through it before it arrives at our eyes. Now if you have ever been in an aeroplane you will be familiar with turbulence, that horrible lumpy bumpy feeling when the aeroplane gets thrown around a little (and sometimes a lot) by moving air currents. These moving bits of air not only disturb the aircraft taking you on your holidays but also beams of light from astronomical objects. The light from stars has to travel through these different packets of air which have different densities leading to the light being refracted or bent by different amounts. The constant bouncing around of the light through the air is what makes the star seem to twinkle.

Interestingly it is only the stars that twinkle. They are so far away from us that they appear in our sky as a point source of light and the atmospheric effects are much more pronounced. The height of a star in the sky, its altitude, also has a greater impact on the amount of scintillation that is seen. The light from stars that are nearer the horizon has to travel through a greater chunk of atmosphere because the incoming light is at a more shallow angle. More atmosphere means more disturbance and a greater degree of scintillation. Planets on the other hand do not twinkle because they are closer and appear in our sky as a tiny disk.

The atmosphere
of Earth. (NASA)

Europa, a moon of Jupiter, could harbour alien life

Our search for life has so far not returned any positive results whether we search deep space or our own cosmological back yard. Mars was once thought to harbour life when early telescopic observations recorded a complex system of lines criss crossing the red planet. It was thought these 'lines' were canals and built by an alien intelligence to transport water across the barren Martian surface. We now know these canals never existed and that they were just an optical illusion caused by poor optical quality of the first telescopes. It is unlikely that we are going to find complex forms of life on Mars, but a better candidate in our Solar System is Europa, one of the Moons of Jupiter.

Europa is one of the four Galilean satellites discovered by Galileo in 1610. With a diameter of 3,120 kilometres it is a little smaller than our own Moon but unlike our Moon it has a water ice crust. Spacecraft that have visited the Jovian system have captured images of linear crack features that could only have formed if the crust was moving independently from the interior of the planet. This discovery along with the general smooth nature of Europa's crust has led us to believe that an ocean of liquid water exists under the icy surface. It is possible that tidal interactions from Jupiter are generating enough heat to melt the ice lower down to create the liquid ocean. It is even thought that the ice is not only melted but it may even be salt water, more like the oceans and seas here on Earth, and it is here where there is a chance we may find complex organisms. We only have to study the environments on Earth where life has evolved to understand that life on Europa may well find a way.

All life on the surface of Earth ultimately relies on light from the Sun to survive. The incoming light gets converted into chemical energy through the process known as photosynthesis in plants. Animals eat the plants and other animals eat those animals, the animals die and help to feed the plants. The entire food chain relies on light from the Sun but at the depths of the ocean no light can penetrate, yet deep at the bottom of the oceans on Earth we find entire ecosystems thriving. Instead these sub aqua eco systems are reliant entirely on the energy pouring out of hydrothermal vents at the bottom of the ocean. Bacteria, tube worms and even some strange breeds of crab scuttle around on the bottom of the ocean with no hope of ever catching a glimmer of light or energy from the Sun above. It stands to reason that if life can survive on the bottom of our oceans then it can just as easily evolve and even thrive on the bottom of the oceans of Europa.

Jupiter's moon Europa
is one candidate for
alien life. (NASA)

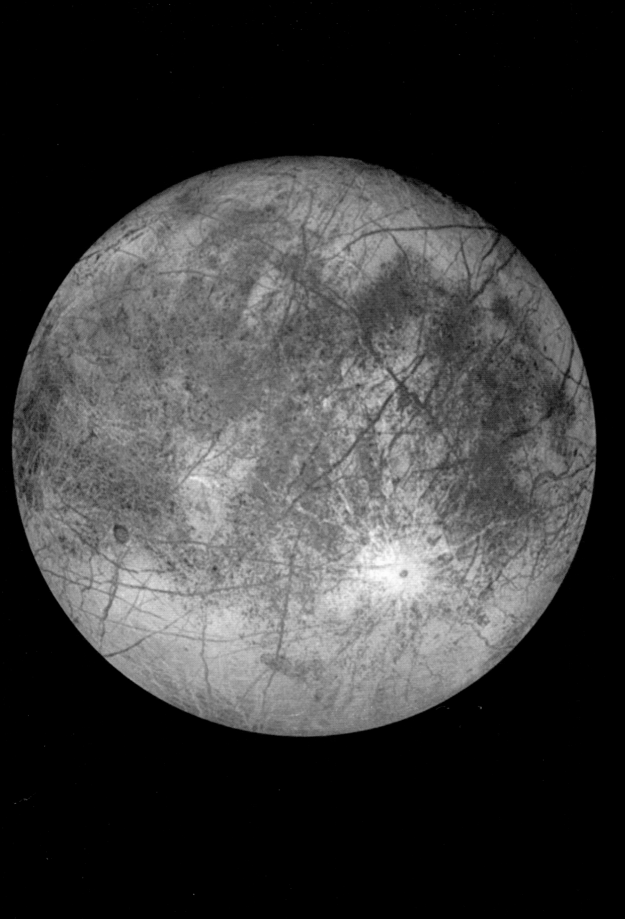

FACT 32

Pluto is no longer a planet

People born before August 2006 will have entered a world with nine planets in our Solar System: Mercury, Venus, Earth, Mars, Jupiter, Saturn, Uranus, Neptune and Pluto – but there was a problem. All the astronomy books revealed nine planets in the Solar System, then we kept finding more, like Sedna, which was discovered on 14 November 2003.

Sedna is just one of a whole list of planetary discoveries that have been slowly increasing the number of known planets in the Solar System. It orbits the Sun at an average distance of 127 billion km (25 times further from the sun than Neptune) and has a diameter of around 1,000 kilometres. Things changed in August 2006 but before then, the International Astronomical Union (IAU) had never defined what something had to be to be classed as a planet.

At the IAU General Assembly in Prague on 24 August 2006, resolution 5A and 6A concerned themselves with the definition of a planet and what that meant for poor old Pluto. The IAU agreed that for anything to be considered a planet it had to comply with the following three points:

1. It must be in orbit around the Sun (or star in the case of exoplanets). Clearly the Earth is in orbit around the Sun, as is Pluto, as is the Moon, as is my car and even me. That means all of us are planets! No, wait...

2. That object must also have 'sufficient mass for its gravity to overcome rigid body forces so that it assumes a hydrostatic equilibrium. In other words, it needs to be roughly spherical in shape. Pluto, the Moon and Earth are still in the running but me and my car are well and truly discounted from planetary status.

3. They must have 'cleared the neighbourhood of their orbit'. Ah now this drops Pluto off the list too.

Applying these three rules to Pluto, in particular rule number three causes Pluto to lose its planetary status for it is joined in its 248-year journey around the Sun by five satellites, the largest of which is called Charon. Charon is 1212 kilometres in diameter, which makes it 51% the size of Pluto, while our own Moon is 27% of the size of Earth. The presence of Charon and its relative size to Pluto means Pluto has not cleared its neighbourhood of other objects and is therefore knocked off the planetary podium.

We now define our Solar System as having eight major planets, five dwarf planets: Ceres (in the asteroid belt), Pluto, Hamea, Makemake and Eris. In addition to these larger items there are a whole host of others that are smaller, less spherical (point 2) and less dominant in their orbits (point 3). The demotion of Pluto caused some discontent in astronomical communities and there were even petitions to get it reinstated. These have all but fizzled out now so it looks like Pluto will now and forever be a dwarf member of our Solar System.

Pluto imaged by the *New Horizons* spacecraft. (NASA)

FACT 33

The first spacewalk nearly ended in tragedy

Space travel is a risky business and it is imperative that those that undertake it are made of the right stuff. This was demonstrated on 30 May 1965 when Alexei Leonov embarked on the first ever space walk. Eighteen months of training were undertaken for this historic activity which came over 4 years before the Apollo Moon landings. The walk was supposed to take place from the Vostok 11 mission but that was cancelled so it was later completed on Voskhod 2. It launched on 18 March 1965 and just a few hours later, Leonov exited the spacecraft and floated around in the cold depths of space. All went according to plan until it was time to re-enter the capsule, then things could not have gone more horribly wrong.

In preparing for his space walk, Leonov had to wear the so called extra-vehicular activity unit which gave him 45 minutes of oxygen while allowing moisture, carbon dioxide and heat to be vented off into space. It took seven minutes to inflate the airlock and he spent just over twelve minutes out on his walk. He recalls the moment he was commanded to get back inside the Voskhod 2 module and it reminded him of the times his mother had called him back in from playing outside with his friends. Unfortunately, getting back inside the airlock posed a little more challenging than returning to his family home.

He had already used 19 of his 45 minutes of oxygen and still had to get back to the airlock and get in. On returning to the airlock he realised that, due to the lack of atmospheric pressure in space, his suit had become very stiff and he was unable to bend sufficiently to get back in through the air lock. He only had one option, to slowly depressurise his suit to give him more flexibility while trying to wriggle back in through the airlock. Depressurising his suit was an incredibly dangerous thing to do but it was Leonov's only option. As he wriggled his way back into the airlock his temperature increased to dangerous levels but eventually he was in. He had to depressurise even further to allow him to bend his body sufficiently to close the airlock door but he made it with minutes of oxygen to spare.

The danger was not over yet though. Just before re-entry it became apparent that the automatic guidance systems had failed and the cosmonauts had to land the spacecraft manually with very little fuel to spare for mistakes. Further problems were experienced when the craft started to spin uncontrollably during re-entry, the cause being the landing module had not fully detached from the orbital module. The spin subjected the craft and crew to 10Gs of force but eventually the cable holding the two modules together burned up and they were able to stabilise and finally land successfully in Siberia.

Astronaut Edward White on the first American spacewalk, just like Alexi Leonov. (NASA)

FACT 34

The Andromeda galaxy is on a collision course with Milky Way

We live in a galaxy called the Milky Way, a spiral galaxy which is 100,000 light years away and thought to be home to 400 million stars. The nearest star to our own Solar System is Proxima Centauri and it lies 4.2 light years away, which is just under the average 5 light years separation between stars in the Milky Way. Between the stars are vast gulfs of interstellar space and gigantic gas clouds and generally speaking this is the case for most galaxies. Between the galaxies it is a little less exciting, with even greater distances of empty space between them. Our nearest major galactic neighbour is the Andromeda Galaxy at a distance of 2.5 million light years away. Unlike more distant galaxies, Andromeda is heading straight for us!

The general movement of galaxies in the Universe is away from us. This is not because we have some form of galactic body odour but that the expansion of space carries galaxies away from each other. The Andromeda galaxy is different however, for it is a member of a group of galaxies known as the Local Group. The Local Group, as its name suggests is a group of galaxies that are local to us in space, and indeed the Milky Way is one of its largest members, along with the Andromeda galaxy and and the Triangulum galaxy. All three of these galaxies are large spiral types, and the Milky Way and Andromeda are host to their own satellite galaxies. Ours are known as the Large and Small Magellanic Clouds and can only be seen from the southern hemisphere. The Local Group is around 10 million light years across and has over 54 members, many of which fall into the category of dwarf galaxies. Many clusters exist like the Local Group and they are all part of larger galactic clusters. Our group is part of the Virgo Supercluster.

The movement of galaxies within a local group is very different to the general movement of galaxies in the Universe. Gravity, one of the fundamental forces of the Universe along with the strong, weak and electromagnetic force, is a relatively weak force but its effects can be felt over vast distances. In the case of clusters of galaxies, the force of gravity is strong enough to overcome the expansion of space. The fate therefore for most groups of galaxies is that they will eventually combine. Our Local Group has this fate waiting for it but not for another 1 trillion years or so. When that time comes, all 54 of our Local Group galaxies will have combined into one super galaxy. This will not happen in one gigantic instantaneous collision but instead, over time, the various galaxies will merge. In the case of the Andromeda galaxy, it will collide with us in around four billion years time. That sounds a long way off but our Solar System is likely to outlive this spectacular event.

The Andromeda galaxy captured by the *Galaxy Evolution Explorer* in ultraviolet light. (NASA)

FACT 35

A Martian volcano is almost three times the size of Mount Everest

There are not many countries that do not have hills and mountains of some sort. The processes that form them are many and varied, from glacial deposits to volcanism. The largest mountain on Earth is Mount Everest in the Himalayas and it rises a staggering 8,848 m above sea level. Around 4,000 people have climbed to the summit of Everest but if they had made their journey a few million years ago it would have been much easier. Everest formed around 60 million years ago when the Indo-Australian plate and the Eurasian plates collided causing them to crumple up, thrusting land up thousands of metres. Everest is merely a bump though when it comes to the rest of the Solar System. Future travellers to Mars can pit their climbing skills against the mighty Olympus Mons, which is nearly three times higher than Everest.

Unlike Everest which is a mountain, Olympus Mons is a volcano and it rises an estimated 25 km above the Martian surface. It has a diameter of 624 km which is about the same size as the state of Arizona, and sports a complex caldera system which is 80 km across. With the height of Mons you might expect it to be visible from hundred miles around as it towers over the surrounding terrain but any future visitors hoping to get a selfie with the volcano in the background are likely to be quite disappointed. Even though its summit is 25 km above ground level it is a shield volcano, which means its slopes rise up very gradually, so much in fact that if you stood at the foot of Mons in certain places, you would barely even know it was there!

Shield volcanos like Olympus Mons get their name from their profile which somewhat resembles a shield lying on the ground. They form over millions of years where highly fluid lava seeps out of volcanic events in the crust. The 'runny' or low viscosity nature of the lava means that over time, sheets of lava accumulate slowly building up the volcano's trademark shape.

Regardless of the sheer size of Olympus Mons its identity had not been confirmed until the space probe Mariner 9 visited Mars in 1971. Before that time, astronomers had detected a high contrast feature at the location of Olympus Mons and had thought it might be clouds wrapped around some as yet unidentified mountain. The feature has since become known as Nix Olympica, which translates to The Snows of Olympus, and it can be seen in amateur telescopes.

Martian volcano Olympus Mons
taken by *Viking 1*. (NASA)

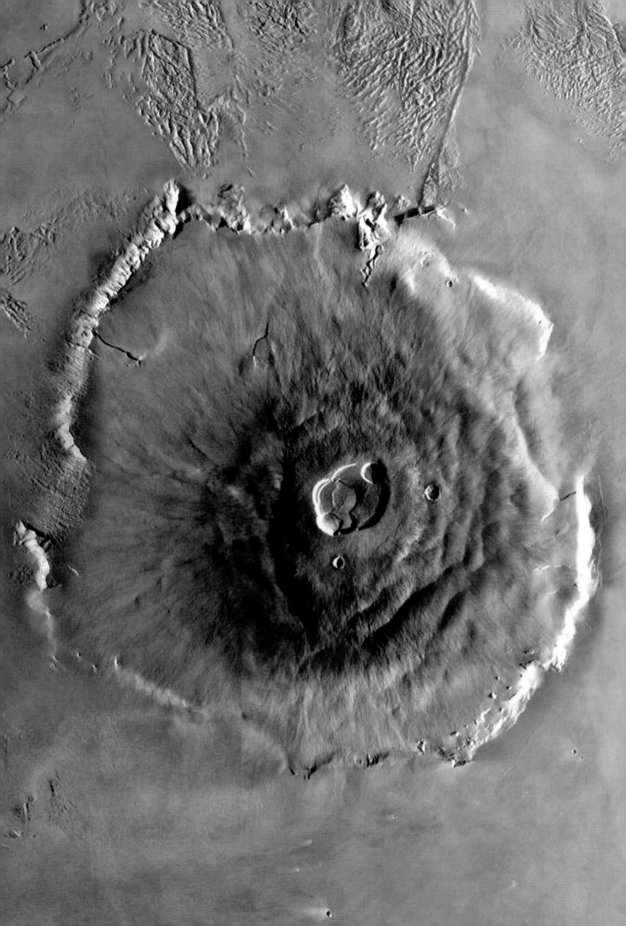

FACT 36

The telescope was invented by an optician

Ask anyone who invented the telescope and you will more than likely get the answer that it was Galileo. It is true that Galileo was the first person to turn the telescope to the sky and in doing so he discovered the the moons of Jupiter, the craters on the Moon and the rings of Saturn. He did not however, invent it. According to records he took the concept that had already been invented and modified it. The generally accepted inventor of the telescope was the Dutch spectacle maker Hans Lippershey, however it is more accurate to say that he was the first to apply for a patent for the telescope.

Lippershey presented the application for his invention in 1608 for a device that could magnify images by three times. This does not sound a lot by modern standards but for the early seventeenth century this was pretty impressive. The instrument had a lens at the front which was convex (outward curve) and an eyepiece lens which was concave (curving inward) and there are many different stories to explain how he got his idea. One story tells of him trying to help a patient with myopia, otherwise known as short sightedness, and in trying to create a pair of corrective 'eyeglasses' he stumbled upon the magnifying properties of aligning two lenses. Another story tells how he got his idea having seen two children in his shop holding up two lenses that made distant objects look bigger. There is even one account that tells of him stealing the idea from Zacharias Jansen, another 'eyeglass' maker in the same town as Lippershey.

Further controversy ensued when yet another Dutchman, Jacob Metius, submitted a patent application for a telescope just a few weeks after Lippershey submitted his. After the applications were reviewed, both were turned down, not simply because there were two applications but more due to the simple nature of the construction and the ease with which they could be reproduced. The invention however, has been generally credited to Lippershey as he was the first to get the patent application submitted.

Galileo came in to the story just a year later when, in 1609 having heard of the new Dutch invention, he set about constructing his own. His design was much the same and consisted of a convex lens and a concave lens. The concave lens was fixed to the eye end of a fixed hollow tube and the convex lens was fitted to the other end. Galileo relied on trial and error to identify the correct separation of the two lenses but eventually identified a configuration that gave him a magnification of 20 times.

The design of telescopes invented by Lippershey and modified by Galileo were the forerunner of today's modern refracting telescopes that use lenses to focus incoming starlight. Unlike their early predecessors, modern refracting telescopes use anything up to three very carefully shaped lenses fitted to a tube that has another, movable, tube connected to it. The moveable tube holds the eyepiece which itself can be made of a number of smaller lenses.

FACT 37

Astronomers use black and white cameras to take colour pictures

Hardly a day goes by when stunning images are released showing the Universe in all its colourful glory. You might be forgiven for thinking that astronomical pictures are taken, well, at the click of a button. The reality is a little more complex and convoluted than that. Astronomers take images of the Universe chiefly for two reasons, the first is for research purposes and the second is simply because they look pretty. The latter reason may seem a little vain, but incredible looking space images can help tremendously during science outreach activities. Images for research purposes can be used to reveal hidden details or enhance certain features depending on the area of research. Certainly the research based imaging can be a little more complex, but even pretty pictures can take quite a lot of effort.

Inside nearly every phone or camera these days is a sensor known as a CCD (Charge Coupled Device) or CMOS (Complimentary Metal Oxide Semiconductor). Which ever device you use, they both do the same thing, namely convert photons of light into electricity. The technology is pretty much the same inside your smartphone but with one difference; an astronomers camera is black and white whereas your smartphone camera is colour. To understand why, we need to delve a little deeper into how they work. The simplest to understand is the mono camera used by astronomers. Think of a mono sensor as a massive great grid of solar panels and when attached to a telescope, the image falls across the grid. Each solar panel will generate a little bit of electricity depending on the amount of light it detects in its portion of the image. The electricity generated by each solar panel can, using a computer, recreate the original image from the telescope. Add in more, smaller solar panels and you will get a more detailed image. Sensors are just like this, they have grids of millions of tiny little solar panels known as pixels which capture photons and convert them to electrical pulses which can be used to re-create the image. The concept is fairly simple: more light means more electricity which translates to a brighter dot formed as part of the image.

Colour sensors are slightly more complex. Instead of a grid of millions of light sensitive pixels, each pixel is composed of pixels sensitive to red, green and blue light and one other to record brightness. This allows a computer to recreate a colour image but the downside is the necessity of having four pixels to capture one tiny little portion of the image instead of one in the mono sensor. The great benefit of mono sensors is that they provide a higher resolution than their colour counterparts.

So that we can still get glorious colour pictures and retain the high resolution given by mono sensors, astronomers generally take a mono picture through a red filter, then another image through a green filter and another through a blue filter. An additional high resolution mono (black and white) image is also captured and all four are then recombined through image processing to give a colour picture. Astronomers can go further with this and use different filters with mono sensors such as hydrogen alpha to capture the light emitted by hydrogen gas clouds. The mono camera gives much more flexibility but the downside is the additional time it takes to grab all the images to show how beautiful the Universe is.

FACT 38

There are thousands of planets beyond our Solar System

Most people are taught in school that there are eight planets in the Solar System from innermost Mercury to the most distant planet Neptune. This was indeed the case until 1992 when several exoplanets were discovered orbiting around a star with the rather catchy name of PSR B1257+12. In the years that followed more and more exoplanets were discovered until today's tally reaches in excess of 4,000.

   Those planets found orbiting other stars are found in a number of different ways. One of the easiest to understand relates to the study of light from

Fomalhaut b is an exoplanet 25.1 light years away. (NASA)

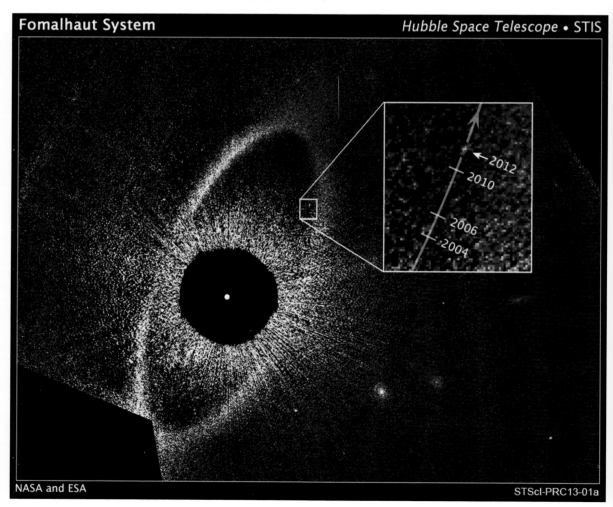

Fomalhaut System   Hubble Space Telescope • STIS

2012
2010
2006
2004

NASA and ESA   STScI-PRC13-01a

distant stars. In its simplest form, photometric studies allow us to monitor the brightness of distant stars. A tiny dip in brightness may be the result of the star being a variable star but it is also possible that the reduction in light is caused by the passage of a planet between us and the star. By studying the light curve of distant stars we can identify those which have planets in orbit around them.

Another method for finding planets around distant stars is to study their position. To understand how this works, it is useful to understand that, for example, the Moon does not orbit around the Earth, but instead it orbits around the system's common centre of gravity. A distant alien observer could study they Earth, perhaps detect it is wobbling a little and be able to calculate the mass and distance of the Moon. If you have a very massive object and a less massive object, then that centre of gravity will be much nearer to the centre of the more massive object. For two objects of identical mass, the centre of gravity would be directly inbetween them both. Astronomers can look for stars whose position shifts slightly, the tell tale signs of a wobble caused by a planet. For stars that are too distant or for planets that are smaller, then its likely that the wobble cannot be detected visually. In these cases the wobble may be detectable by looking at the spectrum of the star. If the star is wobbling due to an unseen planetary companion then it may show up as a regular red shift followed by blue shift in the spectrum. The red and blue shift of various spectral features is similar to the sound of an emergency vehicle rushing towards you making the pitch of the sound increase then decrease as it heads away again. The movement of the star as it wobbles around the centre of gravity can be detected in this way.

At the time of writing, over 4,000 exoplanets have been found orbiting around distant stars. The method of detection of these new worlds is, as you have seen, really quite varied. Regardless of how they were found, it clearly shows that planetary formation is common in our Universe.

FACT 39

There is a hurricane on Jupiter nearly three times the size of Earth

When Galileo turned his first primitive telescope on mighty Jupiter and discovered the four Galilean satellites little did he realise that he was observing the largest planet in the Solar System. As telescopes improved, astronomers after him started to detect finer details. In 1664, English philosopher and astronomer Robert Hooke was likely among the first to observe the large 'permanent' red spot on Jupiter, although it is possible that it was observed earlier. The true origin of this rather innocuous blemish in the clouds of Jupiter was not truly understood until the observations from the Voyager spacecraft which passed the planet in February of 1979.

Jupiter, like the other outer planets is a gas giant, and has a diameter of 139,822 kilometres. The Great Red Spot, as it has become known, is a colossal hurricane system just like those here on Earth but almost three times as large as our planet! It rotates in a counter-clockwise direction just like the low pressure systems here on Earth but takes six Earth days to complete one revolution. With wind speeds of over 600 kilometres per second it not only dwarfs hurricanes on Earth but due to its oval nature it measures 30,000 kilometres across by 13,000 kilometres wide, making it so large it could swallow up nearly three planet Earths.

Those who have observed Jupiter's Great Red Spot will know that it always seems to keep at the same latitude, located in the Southern Equatorial Belt. It has remained stable here for centuries, only varying in longitude as it drifts around the planet. Estimates suggest that it has completed ten laps around Jupiter since the early 1800s. Curiously the spot has been around for hundreds of years but like all hurricane systems it should be reducing in size; however recent studies seem to indicate it is shrinking at a slower rate than expected. Perhaps the hurricane is drawing energy from deeper in the Jovian atmosphere and it is this which is keeping it fuelled and slowing its demise.

One of the greater mysteries is the cause of the hurricane's red colour. There are many theories that try to explain this and they range from the presence of complex organic molecules to the presence of sulphur compounds that have a distinctive red colour. One thing we do know about the colour is that it is not constant. The spot can vary tremendously from a deep red through to pale pink and even white. When the spot turns white it seems to vanish from view but there is an interesting correlation between the appearance of the southern equatorial belt and the Great Red Spot. When the southern equatorial belt is dark, the spot seems to be particularly light but when the belt is light, the spot seems dark. For now, there is no satisfying theory to explain the colour, let alone the changing colour of the Great Red Spot which, even 400 years after its discovery, could still swallow up the Earth.

Jupiter's Great Red Spot, a hurricane nearly three times the size of Earth. (NASA)

If you fell into a black hole you would be stretched like spaghetti

As we saw in fact number 18, black holes are not actually holes in the fabric of space. Instead they are the result of the collapse of a massive star at the end of its life. The reality is that we have never 'seen' a black hole, but we have seen phenomena that conform to the hypothesised characteristics of a black hole, for example the intense radiation emissions from gas as it gets sucked in or the red shift of light from material being dragged around the event horizon. We can observe these effects but what we cannot study is what would happen if we fell into a black hole.

To understand what is likely to happen it is useful to have an understanding of their anatomy. The powerhouse of the black hole is something known as a singularity. The singularity is the stellar corpse that has been compressed to such an extent that it now has infinite density. We have already seen that objects with mass curve the fabric of space-time, with the more mass the higher the curvature. Singularities curve space-time to extreme levels and if you were to get to one you would find the known laws that govern space and time all cease to be relevant.

Due to the extreme curvature of space-time the gravitational pull from a singularity is higher than anything experienced anywhere else in the Universe. If you want to escape from the gravity of Earth for example then you need to travel at 11.2 kilometres per second, but to escape from a singularity requires speeds in excess of the speed of light, faster than 300,000 kilometres per second! An important point to consider now is that any event in the Universe communicates itself to the surrounding space through electromagnetic radiation. For example a stellar flare might send out a burst of X-rays or giant gas clouds might gently emit radio waves. For that reason, there is a region surrounding a singularity known as the event horizon and if something exists within the event horizon then there is no way that the

The blackhole at the centre of galaxy NGC1068 is revealed thanks to the high energy x-rays. (NASA)

outside Universe can ever know anything about it since any electromagnetic radiation (such as light) being emitted can never escape. Event horizons can be anything from tens of kilometres from the singularity to hundreds and even thousands of kilometres away for supermassive black holes.

If you were to ever be unlucky enough to be sucked in to a black hole then an observer watching you fall through the event horizon would see you travel slower and slower before seeming to freeze at the point of no return at the horizon. Your image would be frozen in time, but for you falling 'into' the black hole your body would be subject to the immense pull of gravity. Due to the nature of gravity and the effects of the singularity the gravity on your toes would be stronger than the gravity on your head so your body would be stretched out just like spaghetti. This process even has a name: spaghettification.

FACT 41

There are clouds of water floating in space

Take a look at images of planet Earth from space and its easy to see our planet is covered in water; in fact about 71% of the surface is water. Looking up at the dark night sky you might think water can only be found here but studies of bodies in our Solar System reveal water in various forms on the Moon, Mars, Enceladus, Europa and even comets. It seems water is fairly common in the Solar System but as we explore deeper into the Universe we are even finding it in remote corners of the cosmos.

Quasars are galaxies that have enormous black holes at their core. In the night sky they appear to be very small objects at huge distances from us, emitting vast amounts of energy. Studies of a quasar with the rather catchy name APM 08279+5255 have revealed something rather surprising. It has a black hole in its core that is estimated to be 20 billion times mores massive than the Sun and lies at a distance of 12 billion light years from us. The black hole is feeding on surrounding clouds of gas and dust which has caused it to grow to extreme sizes. In 2011, a discovery announced that not only did this massive black hole have the usual disk of material spiralling into it but there was also a huge cloud of water vapour, so large that it is estimated that 140 trillion times more water exists than is in all the oceans on Earth.

The water vapor around APM 08279+5255 has been found to be distributed in a gas region around the black hole that extends for hundreds of light years. Intense X-rays and infrared energy from the quasar are bathing the cloud, warming it to a toasty -53 degrees! Sounds quite chilly but in the depths of space, this is positively tropical. We can estimate the amount of gas and water vapour in the cloud and current theories of black hole evolution suggest there is enough material to keep feeding the black hole until it is six times larger than it is today.

The study that led to this discovery started in 2008 when a team of astronomers used a telescope that has a 10 metre diameter at the summit of Mauna Kea in Hawaii. The study was then later enhanced with follow up observations from the CARMA (Combined Array for Research in Millimeter-Wave Astronomy) array of radio telescopes in South California.

Artists impression of APM 08279+5255 which contains water vapour. (NASA)

FACT 42

It is very likely that the Sun will one day swallow up the Earth

Death is a natural part of life that most of us accept if not try and put to the back of our minds. What we also tend to accept is that even though our time on this planet is limited, the world will continue to turn and life goes on even if we do not. In fact it is probably true to say that most of us expect the Earth to be around forever, even perhaps longer than humanity itself. We somehow get comfort from the thought that our home planet at least, will survive all of us. Well sadly it turns out that even Earth's place in the Universe could also be limited.

The cause of the demise of the Earth is unlikely to be a planet-smashing goliath of an asteroid that blasts our world into a million pieces but instead our friendly local star. The Sun is a big ball of gas, or plasma to be more accurate, and at the moment it is stable. Nuclear fusion deep in its core is turning hydrogen into

The red giant star Betelgeuse. Our Sun will evolve into a star similar to Betelgeuse. (NASA)

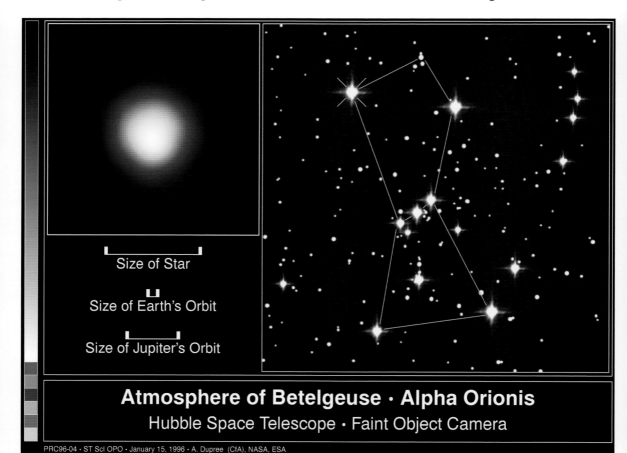

Size of Star

Size of Earth's Orbit

Size of Jupiter's Orbit

Atmosphere of Betelgeuse · Alpha Orionis
Hubble Space Telescope · Faint Object Camera

PRC96-04 · ST ScI OPO · January 15, 1996 · A. Dupree (CfA), NASA, ESA

helium, a process which creates a force known as the thermonuclear pressure. This pressure exerts a pushing force from inside the sun which tries to expand it, while the force of gravity tries to collapse it. Thankfully for us, the two forces balance and the Sun is stable. Over the next few billion years, nuclear fusion in the core speeds up and it starts to emit more energy and that may well mean the end of life on Earth as temperatures increase.

In just five billion years the Sun will enter what is known as the red giant phase. The original hydrogen core will now be full of helium but the temperature and pressure is insufficient for helium fusion to commence. Gravity momentarily wins over the thermonuclear force and the core shrinks. This brings a shell of hydrogen around the core into the fusion zone allowing fusion to start again. With the onset of fusion again the Sun will start to expand and the energy produced in the shell around the core must spread around a much greater surface area, causing its surface temperature to reduce, turning it redder in colour. This marks the onset of the red giant phase of its life.

As the Sun swells its surface is expected to extend beyond the orbit of Earth, which will of course mean the end for our planet. There is hope that it may yet survive though because the expansion of the Sun will also bring with it a loss of mass. As the Sun swells and becomes less massive, the Sun's gravity will become less so the Earth may well slowly migrate out to an orbit further away. Complicating matters further, the tenuous outer layers of the Sun's atmosphere may drag on Earth slowing it down and causing it to drift inwards. Only one thing is for certain, the Sun will expand and get a LOT bigger. Whether Earth gets gobbled up by the Sun or not, no-one knows for sure.

Shooting star over Mt. Fuji at
Motosuko lake, Japan. (Adobe Stock)

FACT 43

Shooting stars are not stars at all

On any night of the year you will quite likely see the rather strange sight of a star or two whizzing across the sky. These 'shooting stars' as they are known are not really anything to do with stars. Stars are so far away from us that their light takes years to reach us so for them to shoot across the sky with the speed observed they would have to be travelling at speeds far greater than the speed of light. The origin of shooting stars is a little more terrestrial than you might think.

When our Solar System formed around five billion years ago it created the Sun and eight planets we see today but the process of planetary formation left us with with a system littered with debris. The pieces of rock and dust hurtling around the Solar System are known as meteoroids or asteroids dependent largely on their size: asteroids are larger, meteoroids are smaller. Their size can range from smaller than a grain of sand to many hundreds of kilometres. Thankfully most asteroids are confined to a belt between Mars and Jupiter but there are countless smaller meteoroids drifting around. If the Earth happens to come across a meteoroid on its journey around the Sun then it will sweep it up like a giant cosmic hoover. As the meteoroid falls into our atmosphere it becomes known as a meteor and something rather wonderful happens.

As the piece of rock plummets earthward it heats up due to frictional forces between its surface and the gas in our atmosphere. The gas surrounding the incoming rock can glow and give off light which we see as the 'shooting star' effect. As it falls into denser atmosphere the effect increases causing the rock to loose material as layers of it vaporise. Most meteors never make it through the atmosphere to land. Some will rain down on the surface as meteoric dust but those that do make it are called meteorites.

These random one-off meteor events are known as sporadic, meaning they are just random lone travellers. Throughout the year though, around 20 times each year, we are treated to meteor showers which are simply the arrival of lots of pieces of debris. When meteor showers peak, usually on a given evening, we will see anything from just a few meteors per hour to hundreds, even thousands for meteor storms. Whilst sporadic meteors are just lone bits of rock, meteor showers are caused by comets as they travel around the Solar System.

Comets are composed of rock and ice, and as they journey around the Sun, they leave a trail of debris, much like a trail of breadcrumbs. If the Earth happens to travel through the orbit of a comet then it sweeps up the debris in its path and we get a meteor shower. Halley's Comet is perhaps one of the most famous comets, and on its 76 year journey around the Sun it sheds material like any other comet. Our orbit takes us through Halley's orbit every October giving us the beautiful Orionid Meteor Shower.

FACT 44

Inhabitants of Saturn could enjoy the northern lights too

As far as we know, Saturn does not have any inhabitants but if it did then they, like us, could look up and marvel at the beauty of the northern lights. The ghostly, eerie lights that flicker and dance in our skies at various times of the year are not just an earthly phenomenon. Indeed the 'aurora', to use their scientific term, have been detected on Mars and Jupiter too. For us Earth-based observers, the aurora that are seen in the northern hemisphere are known as the northern lights or aurora borealis and those seen in the southern hemisphere are known as the southern lights or aurora australis. The displays are usually most visible from sites that are about 10 to 20 degrees away from the north and south poles, but during extreme solar activity they can be seen further towards the equator.

The mechanisms that cause the aurora are the same on all planets and it starts at the Sun. We can see that the Sun gives off light and can feel that it gives off heat, but it also emits something known as the solar wind. These electrically charged particles are released from the Sun's outer atmosphere known as the corona and rush outward, achieving speeds between 250 and 750 kilometres per second. On reaching the Earth the charged particles get channelled around the Earth's magnetic field and have quite a remarkable effect on our polar regions, causing gas in our own atmosphere to glow.

The process is not too dissimilar to that which takes place inside a neon tube. Atoms are composed of a nucleus and any number of electrons. If energy is given to electrons then they gain energy and achieve a 'higher' orbit around their nucleus. Once in the higher energy state they want to get back to their original state and as soon as they do, they emit a little bit of energy. We see that little bit of energy as light and in this case, as the aurora.

If you have ever seen an auroral display then you may have noticed that different colours, subtle but different, can be seen. The different colours are the result of different gasses in the atmosphere and those different gasses have different atomic structure; the atomic structure of hydrogen for example is different from the atomic structure of oxygen or nitrogen. In real terms this means the electrons exist at different energy levels so returning back to their 'natural' state releases different amounts of energy equating to different wavelengths of light and therefore different colours.

The auroral displays on Saturn were detected by the Hubble Space Telescope. The orbiting telescope benefits from being able to observe the Universe in ultraviolet light which ordinarily is absorbed by the atmosphere. Saturnian aurora are not observable in visible light but can only be detected in ultraviolet. This is due to the presence of large amounts of hydrogen in Saturn's atmosphere, unlike in Earth's which is largely nitrogen and oxygen.

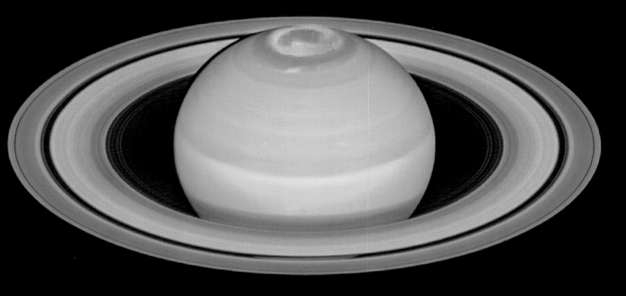

Aurora on Saturn captured by the
Hubble Space Telescope. (NASA)

FACT 45

Comets have two tails

It is believed that the term 'comet' originates from the Greek word kometes meaning 'head with long hair'. The appearance of a bright comet in the sky with its faintly glowing tail may well have given the appearance of a head with long hair to the ancient Greeks, although there is little chance that they genuinely thought it was a head. Throughout the centuries, comets have been seen as an omen of doom, and certainly a comet striking planet Earth could be bad news, but if they stay in the sky then their appearance simply means many late nights for astronomers the world over as they study and observe the almost ghostly visitor.

The great thing about studying comets is that they come to you, or at least some of them swing by Earth on their journey around the Sun. They tend to have highly elliptical orbits and fall into two categories: short period comets whose orbital periods tend to be anything up to 200 years, while long period are anything greater. Halley's comet is perhaps the most well known, with an orbital period of around 75 years. Most comets, even short period ones spend a lot of their time out in the cold depths of the Solar System where ices form on the cometary nucleus. As the comets return to, or venture for the first time into the inner Solar System the warmth from the Sun causes the ice to sublimate. Ordinarily ice melts into liquid and liquid evaporates into a gas but in the case of a comet, the lack of pressure from the vacuum of space causes the ice to turn straight into a gas in the process of sublimation.

Slowly and steadily as the comet's nucleus gets closer to the Sun, more ice sublimates into gas and a vast halo of gas known as the coma forms. It is not unusual for a comet to have a nucleus which is just a few tens of kilometres across and a coma which extends for thousands of kilometres. With the decreasing distance between the Sun and comet, the impact on the latter increases. The ultraviolet component in sunlight strips electrons off the atoms of gas in the coma turning them into ions and the solar wind (the steady stream of charged particles coming from the Sun) carries the ions away from the Sun, forming the gas or ion tail. A second tail is formed as tiny particles the size of cigarette smoke particles get dislodged from the surface, forming the slightly curved dust tail. The pressure from the Sun is responsible for the formation of both tails which can both often be seen quite separate but always both generally pointing away from the Sun. It is perfectly usual for the tails of the comet to precede the nucleus and coma as it heads back out into the depths of space. Eventually the comet's tails and coma dissipate until their next encounter with the Sun.

The Moon was once part of Earth

Think of the night sky and many people will before very long, also think about the Moon. It is so familiar to us that most see it but do not really appreciate it. It is a chunk of rock on average 384,400 kilometres away and 3,474 kilometres in diameter, but to think of it as something we only see in the night sky is wrong. There are many days each month that the Moon can be seen in daylight, but it has not always been visible; moreover there was a time when the Moon did not even exist.

Our Solar System formed 4.6 billion years ago out of a vast cloud of gas and dust known as a nebula. The force of gravity slowly caused the Sun and its family of eight planets to form, but at the time there would also have been a significant number of other large chunks of failed planetary rocks hurtling around. In the young Solar System, collisions were common, indeed we can see evidence of such collisions by looking the surface of Earth. The Gulf of Mexico is thought to be the site of a massive meteorite impact millions of years ago and it is now thought that an impact could have led to the formation of the Moon.

The great impactor theory is just one theory of how the our cosmic companion came to be, another suggested it may have formed at the same time as the Earth as a companion. Geological studies of the Moon show that it is less dense than Earth and had they formed at the same time then they would have had broadly similar composition, but the Moon has less heavy elements than Earth. Another theory suggest it was a distant Solar System cousin that simply wandered too close and got captured by the Earth's gravity. We can see examples of this in other parts of the Solar System and typically captured objects are oddly shaped and their orbits are usually not so well aligned to the Earth's path around the Sun. These ideas have largely been discounted in favour of the great impactor theory.

The great impactor theory suggests that an object, which has been named Theia, smashed into the Earth around 100 million years after the Earth formed. This Mars-sized object cannonballed into the Earth sending chunks of crust into space. Over time, the force of gravity acted upon all of these bits of rock in orbit around the Earth and slowly bound them together into the Moon we see today. This model explains how the Moon is less dense than the Earth and explains why the Moon is composed of material that is like the rocks forming the crusts of Earth. Impacts of this nature can spell global catastrophe and wipe out entire populations but without the impact of Theia we may have never been able to gaze upon a moonlit scene on a clear night.

The Moon.
(Mark Thompson)

FACT 47

To find black holes, astronomers look for the brightest objects in the night sky

A black hole is a massive star that has reached the end of its life. The term black hole comes from the extreme gravitational field surrounding them that stops even light escaping. Light or more accurately electromagnetic radiation travels through a vacuum at 300,000 kilometres per second, and even at that speed it cannot escape the pull of the black hole. There is a defined distance known as the event horizon within which light cannot escape. Given that light cannot escape, it might be fair to assume nothing can escape, yet astronomers have discovered that the best way to find them is to hunt among the brightest objects in the Universe.

Objects and material that get pulled towards a black hole but do not fall in tend to form a spinning, flattened disk known as the accretion disk. Any particles getting dragged into the accretion disk will start bouncing against other particles and through frictional heating, lose energy. As the particles lose energy, they try to move into a lower orbit but as they move into a lower orbit the immense gravitational pull from the black hole accelerates them to a higher velocity. Particles that migrate towards the event horizon will have lost energy through frictional heating but will have gained velocity through gravitational acceleration.

The material in the disk can get accelerated to 'astronomical' speeds and as it falls ever closer it gets compressed. The combination of this and frictional forces within the disk cause the disk to emit energy in the form of electromagnetic energy of varying wavelengths. By the time the material has reached the proximity of the event horizon, it has enough energy to start emitting X-rays, some of the most energetic forms of electromagnetic radiation, and causing it to shine out like a beacon. These beacons are known as quasars and their energy output can outshine an entire galaxy of billions of stars.

The spinning accretion disk can have another quite surprising effect on the most massive black holes. The disk material will often have electrically charged particles in it and they will bring along a powerful magnetic field. As the material gets close to the event horizon, the theory suggests that the drag from the faster spinning black hole (assuming this is of course a spinning black hole) will wind the magnetic field into a growing cone along its rotational axis which turns into a jet. The jet can then propel disk material out from the poles of the black hole. This is just a theory for now and requires further studies on the rotational speeds of supermassive black holes.

FACT 48

There are three laws which govern the motion of the planets

Io
Callisto's shadow
Callisto
Europa's shadow
Europa
07:10 UT

Jupiter and its moons are governed by Kepler's laws of planetary motion. (NASA)

In the sixteenth century, Nicolaus Copernicus shocked the world when he published his Sun-centred model of the Solar System. This was hugely unpopular with the Church, which believed that the Earth was at the centre of everything. Johannes Kepler was a German mathematician, who in 1600 met astronomer Tycho Brahe. Brahe had accumulated observations of Mars, and Kepler set about analysing them. Brahe was impressed with Kepler's analytical skills and eventually they entered into a financial arrangement for Kepler to work for Brahe and continue to analyse the observations. As a result of his work with Brahe's observations, he published his three laws of planetary motion in the early 1600s.

The first law explained that all planets orbit the Sun in elliptical orbits, with the Sun at one of the foci. The Copernican heliocentric system suggested planets orbited the Sun in complex circular patterns but Kepler simplified this by suggesting in his first law that the orbits of the planets are elliptical. An ellipse is just a squashed circle with the amount of 'squashedness' being defined by the term known as eccentricity. An ellipse with an eccentricity of 0 is a circle and as the number increases, the circle gets more squashed. Kepler described the orbits of planets as ellipses, which negated the need for the complex circular orbits and system of epicycles and deferents developing in the Copernican system.

Kepler's second law states that an imaginary line joining the planet to the Sun, known as the radius vector, sweeps out equal areas with equal periods of time. With a circular orbit a planet can follow a regular speed but this was not observed. Instead, the elliptical nature of planets orbits brings them closer to, and further away from the Sun. If the planets orbital speed were constant then the radius vector would not sweep out equal areas in an elliptical orbit. Instead, as the planet gets closer to the Sun, it speeds up and as it gets further away, it slows down. This changing velocity in the orbit allows the radius vector to sweep out equal areas of space. When further away, a longer thinner slice is carved out whilst a shorter but wider slice is traced out when nearer the Sun and moving faster.

Finally the third law of motion states that the square of the sidereal period of a planet is directly proportional to the cube of its mean distance from the Sun. This law could be changed so the last part could be read '...mean distance from the object it orbits' because it can be used equally well for moons in orbit around their parent planet. This is a really useful formula because it means if we can determine the time it takes for a planet (or moon) to orbit the Sun (or its planet) then we can calculate its distance from that object, or conversely if we know how far away it is, we can calculate the time it takes to complete an orbit.

The most common star is a red dwarf

It is not obvious when you look at a star-filled sky that the stars have colour. Look closely though and you will soon notice that you can start to see colours: yellow, white and blue, but if you look at the right time of year you might be able to spot a red star or two. Stars vary in colour but they vary in type too: there are yellow dwarfs, brown dwarfs, blue giants and even red supergiants. If you surveyed all the stars in the night sky and grouped them by type then you would find the biggest group by far would be the red dwarfs.

Red dwarfs, as their name suggests are red in colour and small in size but their low luminosity means they are impossible to detect by the naked eye alone. Telescopic surveys of stars in our region of the Milky Way suggest they are the most common stars in the Galaxy. In fact it is thought that three quarters of the stars in the Milky Way are red dwarfs and of the 60 nearest stars, 50 of them are red dwarfs. The nearest red dwarf star to our own Solar System is Proxima Centauri, part of a multiple star system which is just over four light years away.

If you want to find an exact definition for a red dwarf you will struggle to find one. They are stars that form in the same way as any other star, out of a cloud of gas and dust. Gravity slowly takes hold and causes the cloud to collapse and that brings with it nuclear fusion, which marks the birth of a star. A red dwarf star will typically be born with a mass less than 50% of the mass of the Sun. The lower mass means that they are cooler than other stars (by comparison a red dwarf is around 3,500 degrees C compared to the Sun at 5,500 degrees C) and therefore burn through their fuel at a slower rate. More massive stars only burn through the hydrogen in their core before evolving further while red dwarfs burn through every last gram of hydrogen in their body.

It is often difficult to differentiate between red dwarfs and their cooler cousins the brown dwarfs. A brown dwarf is an object that never quite gained enough mass to start the nuclear fusion process, in effect they are stars that never ignited. As a result, they are cooler than other less massive objects. Spectral studies of these objects reveals their temperature and also the hidden processes deep in the star.

Many red dwarf stars have been found to have planetary companions which formed out of the leftover disk from the formation of the star. Any planets around them are considered to be unsuitable for life to evolve. The parent star's low light and heat means the zones within which planets could support life would be very close to the star itself. Being so close sounds great but the radiation from the star will be too high for most known forms of life to evolve.

The red dwarf star Gliese 623 showing one of its accompanying planets 623b. (NASA)

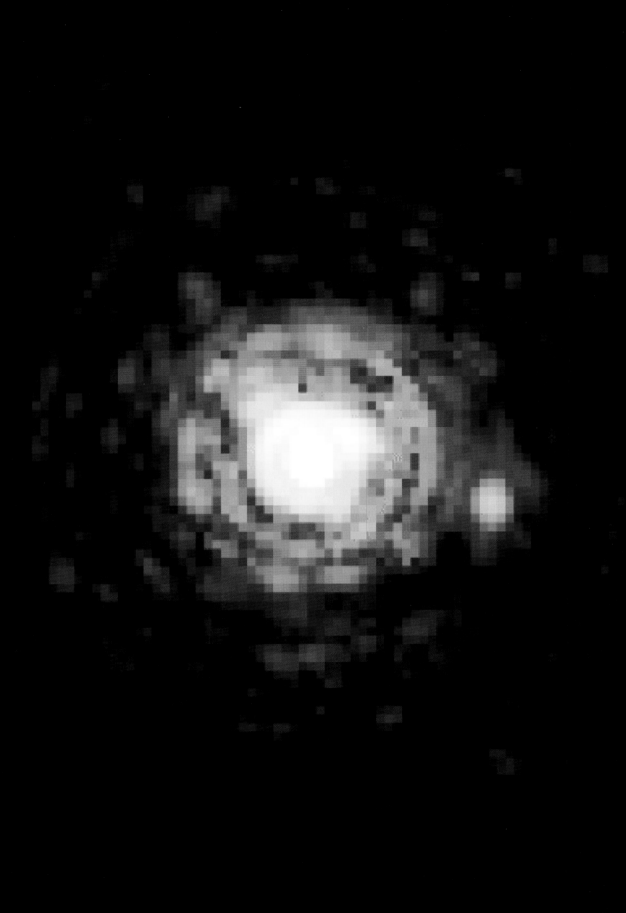

FACT 50

The Kessler Effect suggests one day we may get cut off from space!

The Kessler effect was proposed by NASA scientist Donald Kessler in 1978 and suggests there is a strong likelihood that the quantity of objects in low Earth orbit will increase due to collisions which could cascade out of control. If this happens then the 'minefield' of projectiles whizzing around our planet would simply make a journey through it far too dangerous to undertake.

Figures from 2015 reveal there were over 2,000 manmade satellites orbiting the Earth and an estimated 70,000 pieces of junk floating around that range in size from 1 centimetre to 10 centimetres. It is likely that there are hundreds of thousands of pieces smaller than this size that are not easily detected or tracked yet still travelling fast enough to burst a hole in an unsuspecting astronauts space suit.

With all those objects in orbit it is not surprising that at least one satellite per year is destroyed by space debris. It has happened on more than one occasion that the occupants of the International Space Station have had to wait in their escape module while a piece of space debris safely passes by. It is difficult to know if an impact is likely so it is better to be safe and take precautions.

Kessler suggests that the destruction of a satellite would lead to much debris and it is here that the danger lurks. Each piece of debris from a destroyed satellite would pose a threat to other satellites. If a satellite were to break into 10 pieces, then the destruction of the first could lead to the subsequent destruction of 10 more and these 10 could lead to the destruction of 100 and so it goes on.

Thankfully space agencies require mission designers to take this all seriously and any new spacecraft is required to demonstrate that it can be safely disposed of at the end of its life. This can either be done through a deorbital burn to return the craft to Earth or burn it up in the denser atmosphere below. An alternative is for the disused spacecraft to be moved to a graveyard orbit which are orbits well away from any other commercial operations.

As things stand, we are not currently cut off from space but perhaps we are starting to see the tip of the iceberg. If we act quickly we can stop making matters worse but proposals are underway that will perhaps start to improve the situation. Ideas for clearing up space have ranged from the slightly crazy to the downright absurd. Ideas like massive sheets of nothing more than what can only be described as sticky cosmic fly paper or nets used to ensnare debris. More serious ideas include the laser broom. This land-based laser would be directed at orbiting debris between 1cm and 10cm and its energy would cause part of the debris to vaporise. The vaporisation would cause the emission of a little energy to act as a tiny little rocket engine and provide thrust to deorbit the debris.

The International Space Station has been evacuated on a number of occasions due to potential debris impact putting the inhabitants at risk. (NASA)

FACT 51

The first telescope mirrors used arsenic

The first astronomical telescopes used lenses to focus incoming starlight but the original designs suffered from an optical effect known as chromatic aberration. The incoming light was a mixture of all 'colours of the rainbow' but their different wavelengths meant the simple telescope lens focussed them at different points. An astronomer looking through one of these telescopes would see an image surrounded by colour halos as the different components of light focussed at

The mirror of the Hubble Telescope being made. (NASA)

different points. In the seventeenth century, Isaac Newton was credited as the first person to develop a telescope using spherical curved mirrors instead of lenses to remove the unwanted effects of chromatic aberration. This was largely successful but other unwanted optical effects plagued the reflecting telescope as they became known.

Newton's first telescope, which he built in 1668 used a hitherto groundbreaking design around a 3.3cm circular metallic mirror. The mirror had a curve manually ground into its front surface such that its shape could have been formed by pressing it onto a sphere. The mirror was fixed mechanically to the bottom of a tube such that starlight would fall down the tube and strike the curved surface. The light was then reflected back up the tube but the curve of the mirror meant that the reflected light was in the shape of a cone as the incoming light was being brought to a focus at the point of the cone. Before the light came into focus, it would strike a smaller, flat 'secondary' mirror that redirected the cone of light out the side of the tube where the astronomer would be able to observe the object through an eyepiece. This design known as the Newtonian reflector became one of the most popular telescope designs of all time for both amateur and professional astronomers.

The metallic mirrors in the earliest reflecting telescopes were made of a pretty gruesome mixture of two parts copper, one part tin and a little bit of arsenic. Later concoctions removed the need for arsenic and often included lead, zinc and even silver. Whilst the mixture provided greater reflectivity than other designs (reflecting 66% of light that hit the mirror compared to modern professional instruments reflecting up to 95% of the incoming light) they tarnished rapidly when exposed to air so they had to be regularly re-polished to maintain their reflectivity.

Things changed in the mid 1800s when a new process was developed to deposit ultra-thin coatings of silver onto glass. Glass was much easier to work with, held its shape better and the use of silver coatings meant the telescope mirrors required much less maintenance. This was a vast improvement on earlier techniques, and certainly the removal of arsenic was welcomed. Being a telescope mirror maker is now a pretty safe job but was dangerous when arsenic was involved. With effects like skin colour changes, vomiting, severe nervous system damage and even cancer, life expectancy was quite short!

FACT 52

A gas cloud contains enough alcohol for everyone for a billion years

Gas clouds in space are known as nebulae and they are regions out of which stars form. The clouds often contain gas including hydrogen, other elements and tiny pieces of dust. Over millions of years, electrostatic forces lead to the earliest stages of stellar formation; indeed our Sun formed out of one such cloud 4.6 billion years ago. Nebulae are found all around our Galaxy and in other, distant galaxies but the study of one particular cloud in the Milky Way is enough to raise the eye of any beer-loving individual.

In 1995 a team of British astronomers using the James Clerk Maxwell telescope in Hawaii announced the discovery of a new cloud at a distance of 10,000 light years in the constellation of Aquila. It has the rather catchy name of G34.3 and is a whopping 1,000 times larger than our Solar System, and at its centre is a hot young star. Within the cloud is ethyl alcohol or ethanol, the kind of alcohol that gives your weekend drinks that extra special ingredient. Ethanol is just one type of alcohol but it is usually the other types, such as methanol, that are detected in space.

The processes that lead to the formation of ethanol are usually biological in nature and rely on the formation of sugars. The metabolic processes in yeast can produce the necessary sugars as can overripe fruit. In the depths of interstellar space, biological processes are unlikely to exist and instead it is now thought that dust particles in gas clouds could provide the much needed safe haven. It is just possible that molecules gather on the dust particles and allow for the complex chemical interactions for alcohol to form. Over time, the ethyl alcohol forms on the dust grains, and as it drifts close to the star at the centre of the cloud, the molecules warm and evaporate turning into gas. The gentle evaporation is sufficient to not destroy the delicate bonds between the molecules leaving them to drift through the cloud as molecules of quaffable gas.

Ethyl alcohol has been detected in space before, such as in the Sagittarius B2 molecular cloud near the centre of our Galaxy, but it seems that the alcohol in G34.3 is enough to dwarf any other discoveries to date. One pint of beer usually contains about 5% of alcohol by volume. By studying G34.3 spectroscopically it seems that it contains enough alcohol to produce 300,000 pints of beer for every person on Earth, every day for a billion years. Now if you need a reason to support space exploration, that is your reason right there.

A gas cloud in the Large Magellanic Cloud similar to those found with alcohol inside. (NASA)

FACT 53

Our eyes are rubbish colour detectors in the dark

It is a common occurrence for newspapers, magazines and social media feeds to be full of wonderful coloured images of the Universe. Star clusters, spiral galaxies peppered with coloured stars and incredible glowing clouds of all imaginable colours are inspiring to say the least, yet as astronomers, we need to be careful to articulate that, whilst these images are often true colours, if you excitedly turn a telescope to the heavens to see them for yourself then you need to be prepared for a disappointment. Cameras are great at detecting colours when there is not a lot of light about, but our eyes alas, are not. Turn your telescope on the night sky and take a peek at that beautiful gas cloud only to see a grey/green faint fuzzy blob.

We can thank the evolution of our eyes for this underwhelming performance. The business end of our eyes is the retina and upon it are millions of tiny little light-sensitive detectors. There are two types of detectors that perform very different jobs: the cones which are active under high levels of light such as daylight or in an illuminated room at night, and the rods which kick in when light levels are reduced, such as during night time.

There are six million cones on the retina and among their number, some detect red light, some green and others blue. They work together to give us the colour view of the world we see, and due to their higher density we see a nicely detailed view too. When we are in a brightly illuminated environment, the cones are happily working away, but if you find yourself suddenly in a darkened environment the eye has to adapt. The cones are the first to respond and become a little more sensitive after just a few minutes but the rods take longer to kick into action and respond. Rods, of which there are about 100 million, take about 30 minutes to adapt to the darkened environment but when they do, it is these that are far more sensitive.

When it comes to a darkened environment, rods are the detectors you want because they are much more capable of detecting subtle changes in light. Imagine being a cave-dwelling ancestor out at night; it is far more important to see that a lion is moving around ahead of you than it is to see what colour it is! For that reason and to ensure our very survival, evolution has given us eyes that are not the best for studying the Universe but they will give us the best chance of staying alive. When an astronomer walks out of an illuminated room, it takes about 30–40 minutes for their eyes to become fully adapted to the darkened environment, which means their rods are then at their highest sensitivity. The view that then greets them through a telescope is one constructed by the colour-poor operation of the rods. Thankfully, cameras can detect colour to give us a rather more spectacular view of the cosmos.

Light takes around one million years to reach us from the core of the Sun

Sat in the sunshine on a summer afternoon and your thoughts may drift off to your next summer holiday or that the grass needs cutting, but you probably do not think too much about the journey of a photon of light to get to you. Light is generated in the core of the Sun as a byproduct of the fusion process. Currently hydrogen atoms are smashing together to form helium atoms and in the process a tiny amount of matter is converted into helium in accordance with Einstein's equation E=mc2. The journey from that point on is not quite as straightforward as you might expect.

The first challenge facing a fresh-faced photon of light that has just been released is the sheer density of the material inside the Sun. The atoms in the core are mostly hydrogen atoms that have been stripped of their electrons. They are packed so tightly that there are a thousand trillion trillion atoms in one cubic centimetre and the mean free path (the average distance travelled by a photon before it hits another atom) is approximately 1 centimetre. Photons inside the Sun travel this distance in about 30 billionths of a second, but another consideration must be taken into account: the photon has no idea which direction 'up' is or in other words, it just randomly bounces around from one atom to the next not really knowing which direction it should be going.

A great analogy for this is to think of a drunk trying to walk home. On the way to the pub, the drunk was quite sober and walked in a straight line, perhaps taking just 30 minutes. On the way home and full of alcohol the drunk is quite likely to keep bumping into things and even walk in the wrong direction at times. The journey will take far longer than it should. In the same way, the photon of light takes a lot longer to escape from inside the Sun than it should do. Estimates range anything from 4,000 years to a million years.

There is no solid surface to the Sun since it is a big ball of electrically charged gas known as plasma. Once the photon pops out of the photosphere, the visible surface of the plasma, then it is on the final leg of its journey. To get this far, the light has only travelled from the core of the Sun to the photosphere, a straight line distance of about 695,500 kilometres, but of course it did not go the most direct route. It has taken thousands, maybe even a million years to travel this short distance but it now has 152 million kilometres of space to traverse. It does this in the somewhat more efficient way of travelling in a straight line, making the journey last just 8 minutes and 20 seconds.

You should never look directly at the Sun so perhaps next time you are sat out in the sunshine, just marvel for a moment at the journey of the million-year-old photons that are gently illuminating the scene.

The photosphere of the Sun where the solar energy finally exits and heads out into the Solar System. (NASA)

FACT 55

A hurricane force wind is just a breeze on Mars

News stories often report hurricanes sweeping across various parts of the world. Even the more mundane 'strong winds' can cause havoc and local disturbances. The phenomenon we know of as wind is driven by the energy from the Sun and it is the same process that drives winds on Earth, Venus, certain moons of Jupiter and even on Mars.

For wind to be experienced on a world there needs to be an atmosphere, as without one, wind does not exist. At its simplest, wind is the movement of gas in our atmosphere. Incoming energy from the Sun warms up the surface of the Earth which as it warms, reradiates the energy at different wavelengths to warm the air in contact with it. The warming air starts to expand, becomes less dense than the surrounding air and starts to rise. This leaves a gap at the surface into which more air flows across the surface to fill the void. It is the movement of air across the surface that we experience as wind. A number of factors determine how strong the wind is but surface pressure is the key factor, although geography and topographical features play some part. The rising and descending of air due to heating from the Sun leads to areas of high pressure (where air is descending) and low pressure (where air is rising) and it is the difference between these that drives the strength of the wind. A greater difference between the two will cause the air to move more rapidly across the surface from areas of high pressure to areas of low pressure.

A wind may be blowing at 60 kilometres per hour but the density of the air that is moving around determines what the wind feels like. Mars is a great example of this difference, and its proximity makes it a fab candidate to study. Telescopic studies of the red planet over the years have often revealed dust storms, giving the illusion of a stormy, windy world. The reality is somewhat different. These continent sized dust storms pop up every year or so and can last for several weeks, blocking out surface detail below and making them conspicuous in even amateur telescopes through the lack of surface detail. Every three to four years these continent-sized storms can whip up into global storms that encircle the entire globe. They sound magnificently scary, but the dusty surface of Mars contains very fine particles so it does not take much wind to lift them up into the air.

Wind speeds on Mars only reach half the strength of the stronger winds on Earth but the Martian atmosphere is only 1% as dense as the atmosphere here on Earth. This means that for a certain wind speed, less air is being dragged around resulting in much less force being exerted upon its surroundings.

The mushroom shaped feature in the centre is a dust storm over the northern polar cap of Mars. (NASA)

FACT 56

An astronaut in space would last about 15 seconds without a spacesuit

Science fiction movies have often majored on the gruesome effects of a human being cast adrift in space without a space suit. Space suits are not needed whilst in the safe confines of your space module because this provides you with all the environmental conditions you need for your fragile body to survive. Remove that, however, and you need your very own portable human-sustaining environment. If for some reason, the door popped off your spacecraft and you found yourself floating in the inky blackness you might be surprised that it is not instant death for you. You would be able to survive for about 15 seconds but after that, the game would well and truly be up.

The main functions of a space suit or indeed spacecraft are to create the right pressurised, oxygenated environment for you to live in, protect you from the extreme heat swings of space and protect you from various forms of radiation exposure. The first effect you are likely to experience is that the moisture on your tongue will start to boil. Between 5–10 seconds after exposure the lack of external pressure would cause the water in the tissues below your skin to vaporise, causing the skin to swell up. Thankfully though, your skin is pretty tough stuff and if you can get back to normal atmospheric quick enough your body would recover.

Soon after exposure, things start to get serious. The lack of oxygen will lead to asphyxiation and you are likely to be unconscious after about 15 seconds as the body uses up all the oxygen left in the blood stream. You might think that holding your breath would help to lengthen this but that would be a very bad idea. The sudden exposure to the lower external pressure would allow the gas in your lungs to expand, rupturing your lungs. The best thing to do if you find yourself exposed to the vacuum of space is to breathe out, releasing some of the pressure in your lungs. If you have your wits about you and you do exhale then you are likely to still be alive after a couple of minutes maybe even permanent damage.

You might think that the sub-zero temperatures in space would be a risk to you and that you may suffer some form of frostbite, but in reality your body would lose heat quite slowly. Of more concern would be the exposure to radiation, and you might pick up some nasty sunburn too. Certainly adrift in space is not a good place to find yourself and whilst it does not mean instant death, the prospects are not great for you. Any rescue needs to be swift but you do stand a chance which is greater than zero.

Buzz Aldrin stood on the surface of the Moon, protected by his space suit. (NASA)

Travel fast to stay young

For hundreds of years, humans have been searching for the magical elixir of life. Some expect it will halt the signs of ageing, others perhaps more ambitiously hope that it will halt ageing itself. You would be hard pushed to find a scientist who actually believes the elixir exists, yet even Einstein made a prediction that might just prove to be an elixir all of its own. According to special relativity, space and time are two sides of the same coin. You cannot have space without time, and indeed you cannot change space without also having an effect on time.

This relationship between space and time is easy to visualise if you imagine a meeting with a friend. You might agree to meet at 7pm on Thursday at the pub called Photon's Return, which is at No.1 The High Street. The address gives you the co-ordinates in space but you need another co-ordinate to successfully identify the meeting place, you need the time! You could arrive at Photon's Return but at 8pm on Wednesday, not Thursday, but in this case you are at the right location in space but the wrong location in time. To identify the location of an event in the Universe we need to know where it is in both space and time. Understanding the nature of the intricate relationship between space and time leads us to the conclusion that you cannot have space without time, nor time without space.

Special relativity goes further and demonstrates how changing one has an effect on the other. For example, travel through space, and time gets affected. It goes on to show how two clocks run at different rates if they are moving at different speeds. For example, put a clock on a space ship and fly off at high speed and any clock on board will 'tick slower' than a similar clock stationary on Earth. This effect is known as time dilation and it is this that can perhaps unlock science's version of the elixir of life.

Time dilation is not just a theory, it has been tested in practice. Atomic clocks provide for incredibly accurate time keeping and they have been used to prove that time dilation exists. A pair of atomic clocks have been used to study its effects, and experiments have shown that one sent on a high speed journey ran slightly slower than its stationary counterpart. In other applications, such as clocks on board GPS satellites, further evidence supports how relative speed has a very real and measurable impact on the relative passage of time. Indeed if the effect of time dilation were not corrected for in GPS systems then they would be far more unreliable and many more people would find themselves very lost!

Perhaps the most exciting impact of time dilation is the prospects for slowing down the effects of ageing. Book a seat in a super-fast craft, the faster the better, and head off on a holiday. By the time you return, your identical twin will have aged a lot more than you, or in other words, you will have slowed down the ageing process relative to anyone here on Earth during your journey.

FACT 58

The sky has co-ordinates just like latitude and longitude

Look at a map of the Earth and you will see a grid pattern criss crossing the page. The lines are the Earth's co-ordinate system and it can be used to pinpoint the location of any position on our planet. The lines that run across the page are known as lines of latitude and measure position relative to the equator while those running vertically are called lines of longitude and measure position from the Greenwich Meridian. To a newcomer, navigating around the sky can seem somewhat tricky but the sky has a co-ordinate system just like the terrestrial version.

Imagine the Earth surrounded by a massive transparent sphere. This is known as the celestial sphere although of course it is just imaginary. Now imagine the equator being extended out onto the sphere along with the north and south pole. These are known as the celestial equator and celestial north and south pole. The celestial equivalent of latitude is known as declination and starts at the celestial equator which is zero degrees declination. Moving away from the celestial equator is shown as an increasing declination to the poles, referenced by 90 degrees declination for the north celestial pole and -90 degrees for the south celestial pole.

The celestial equivalent to longitude is more tricky to define because we need a fixed point in the sky to start from. Earth spins on its axis so we cannot simply extend the Greenwich Meridian onto the sky, so instead, we use the movement of the Earth around the Sun to define the start point. The ecliptic is the name we give to the plane of the Earth's orbit and it crosses the celestial equator in Virgo and Pisces. This was not always the case, because the Earth wobbles on its axis like a slowing spinning top and as a result the position where ecliptic crosses equator drifts around the sky. When the celestial co-ordinates were defined one of these points was in Aries and to this day is still called the first point of Aries, even though it is in Pisces. The first point of Aries marks the starting point for the celestial longitude system known as right ascension.

Objects in the sky can have their location communicated through this co-ordinate system just like locations on Earth. My home city of Norwich in the United Kingdom has a location of 52 degrees 37 minutes in latitude and 1 degree 17 minutes east of Greenwich Meridian, while objects in the sky have a position, such as the Orion Nebula at -05 degrees 23 minutes of declination and 5 hours 35 minutes in tight ascension. Knowing these co-ordinates an astronomer can pinpoint a location in the sky.

Other co-ordinate systems exist in the sky such as the compass based 'horizontal' system, which measures position based on height above horizon and distance around the horizon from the magnetic north. We have already discussed the 'equatorial' system above and there is also the 'galactic' system based on an object's position relative to the Galactic plane, however, the equatorial and horizontal are those most used by astronomers.

FACT 59

The Grand Canyon is dwarfed by the largest valley in the Solar System

The Grand Canyon is a canyon that was carved by the Colorado River around five million years ago. It is located in Arizona in the United States and measures 446 kilometres long and between 6.4 kilometres and 29 kilometres wide. This is nothing though: spacecraft exploring the planets and moons in our Solar System have found a canyon that dwarfs the Grand Canyon. It has been found on the red planet Mars and is known as Valles Marineris.

Valles Marineris is Latin for Mariner's Valley, since the geological feature was discovered by the Mariner 9 spacecraft which orbited Mars in the early 1970s. The size of the valley makes the Grand Canyon look like a mere crack as it measures over 4,000 kilometres long (approximately 10 times longer) and 200 kilometres wide (over 10 times wider). The valley is so long in fact that it stretches almost a quarter of the way around the planet. There are a number of different theories to explain the formation of Valles Marineris, including erosion from flowing water or from a crack that formed billions of years ago as Mars cooled. One of the current leading theories suggests its formation is related to the formation of the Tharsis Bulge.

The Tharsis Bulge is a large volcanic plateau on the equator in the western hemisphere of Mars and is home to three of Mars' largest volcanoes: Arsia Mons, Pavonis Mons and Ascraeus Mons. It formed through volcanic activity from the three large volcanoes building up the surrounding terrain to the bulge we see today. With the slow build up over millions of years of extra material the crust failed to hold up under its own weight and radial cracks appeared, including Valles Marineris. It is likely that landslides down the steep walls have gradually widened the valley since its original formation.

The valley has a number of 'smaller' geographical features such as the Noctis Labyrinthus on its western edge. In contrast to the structure of the main valley, this region is composed of large cracked rocks surrounded by valleys running in all directions around them. The floor of the region seems to be largely smooth in nature which is thought to be related to the flow of some sort of fluid, perhaps carbon dioxide, lava or even water.

There are a number of outflow regions at various points of the valley such as the Chryse Region to its north. The areas are though to be regions where repeated and catastrophic flooding burst out of the valley. Similar features can be seen on Earth where perhaps ice blocks the flow of water until pressure builds up and the water is catastrophically released. Such events on Earth leave behind a very desolate and barren landscape with a teardrop shape. Many of these features are seen in the Chrysae region.

Valles Marineris shown stretching across the surface of Mars. (NASA)

FACT 60

Earth is closer to the Sun in winter

It is a common misconception that the Earth is closer to the Sun in the summer and further away in the winter. Think about this for a moment though: when we in the UK are basking in the summer Sun, those on the other side of the world are shivering in cold winter weather. When UK inhabitants are huddled around the fire on Christmas Day, our cousins in the southern hemisphere are out partying on the beach. Proximity to the Sun is not so crucial when it comes to defining the warmth of the weather.

The orbit of the Earth around the Sun is elliptical, so at times it is nearer and at other times further away. The Earth is furthest away from the Sun around early July every year at a point in its orbit that we call the aphelion. When it is closest, we say it is at perihelion and this occurs around early January each year, a difference of approximately five million kilometres. As you can see, in the northern hemisphere we experience warmer months when the Earth is further from the Sun so there must be something else that causes the seasons.

Instead of distance from the Sun, the phenomenon that causes our seasons is the tilt of the Earth with respect to the plane of its orbit around the Sun. Imagine the Sun at the centre of a giant sheet of paper with the Earth travelling around it in its elliptical orbit. The giant sheet of paper is known as the ecliptic and it represents the plane of the Earth's orbit. Now imagine the Earth spinning like a spinning top, completing one rotation every 24 hours (although in reality it takes 23 hours, 56 minutes and 4 seconds to spin once). The axis around which the Earth spins is not perfectly upright when measured in reference to the plane of the ecliptic, instead it is tilted by 23.5 degrees from the vertical. It is this tilt which gives us the seasons. When a hemisphere points towards the Sun it experiences summer and when it points away it experiences winter.

When a hemisphere points towards the Sun, the incoming solar energy strikes Earth at a more direct angle but when it is pointing away the angle is shallower. When the angle is shallower, the same amount of energy is spread over a larger area and so more of the ground is warmed. When the angle is more direct it is spread over a smaller area and the heating is more intense. The ground then reradiates the energy to warm the atmosphere which we subsequently feel. The more direct angle of the summer months means the ground gets warmed quicker and we experience warmer weather regardless of our distance from the Sun.

Planet Earth. (NASA)

FACT 61

There is water on the Moon

Water on Earth is obvious for anyone to see, indeed almost 71% of the surface is covered in water. It is not so obvious that water is present in other parts of the Solar System, for example on comets and Mars. Anyone that has studied the Moon through a telescope will attest to the barren nature of the lunar surface, but they may also be surprised to hear that water exists on the surface!

The south pole of the Moon with possible frozen water deposits coloured blue. (NASA)

In 2008, the *Chandrayaan-1* spacecraft was launched by the Indian Space Research Organisation, its destination, the Moon. Previous studies of the Moon hinted at the prospect of water but nothing conclusive had been identified. It is important to note that any water present would be in the form of ice and not liquid. The temperatures are too cold and atmospheric pressure too low for liquid water to exist without evaporating away. Earlier results suggested water ice might exist in the deep dark craters of the Moon's polar regions but the information could be explained by other phenomenon. *Chandrayaan-1* was dispatched to investigate further.

On board the Indian probe was an instrument known as a spectrometer, which takes incoming light and splits it up into its spectrum, much like a prism or a drop of water. Scientists can then study the spectrum from the object under observation and look for signature lines which help us to understand the composition and temperature. Using this instrument to analyse light coming from the deep polar craters as the probe flew over revealed the tell tale signs of ice, water ice. Not only was evidence of water revealed but the spectral lines revealed how molecules of water absorbed infra-red light which differs depending on the water being ice or liquid.

The exciting angle on this discovery was that the ice detected was not buried deep down in the lunar soil but instead, found on the floor of the craters. The orientation of the axis of rotation of the Moon means that there are deep polar craters that never receive sunlight at their floors. In these dark conditions, the temperatures are found to be as low as -156 degrees C. The excitement around surface water ice on the Moon relates to the boost it may give to our manned exploration of the Solar System. Getting explorers and equipment up into space is one thing but the heavier the rocket, the more fuel required and the higher the cost. Trying to launch hundreds if not thousands of gallons of water from Earth is an expensive job. If we can establish a staging post on the Moon which has a lower gravitational pull than Earth then onward exploration, with previously mined lunar water, suddenly becomes significantly cheaper and far more viable.

FACT 62

Dark matter exists but we do not know what it is

Dark matter is dark! That may seem an obvious statement but the very nature of dark matter means it is difficult to study, yet we know it exists or at least it is theorised to exist to account for a number of observational anomalies. The story of dark matter goes back over a hundred years to 1884 when Lord Kelvin (of temperature scale fame) estimated the number of unseen objects in the Galaxy from observations of the movement of stars around the centre of the Milky Way. He constructed an estimate for the mass of the Galaxy based on his measurements and concluded it contained more mass than that observed visually and referred to these objects as 'dark bodies'.

More recent observations reveal that Kelvin may well have been right in his theory of dark bodies or to use today's term, dark matter, pervading the Universe. Similar observations to Kelvin's reveal that the arms of spiral galaxies that rotate around the galactic core do not behave as expected. Stars and gas clouds in orbit around the core of spiral galaxies should orbit at speeds which decrease with increasing distance from the centre but observations reveal quite the opposite. It seems that the rotational velocity of spiral galaxies is constant regardless of distance. By assessing the movement and observed mass in the galaxy we can conclude that there must be significantly more mass surrounding the galaxy that is not detectable: dark matter.

Another observation relates to a prediction by Einstein who suggested that the presence of massive objects could warp space so much that they cause light to seem to bend. The theory was tested during a solar eclipse where distant stars were observed near the limb of the Sun, the presence of which seemed to bend the star light by a tiny amount causing the stars' apparent position to shift. More dramatically, the presence of gigantic galaxy clusters warps space sufficiently that their presence turns them into space busting gravitational lenses, bending light from more distant galaxies. Observation tells us how much material we can see in the intervening galaxy cluster but studying the 'lensed' image of distant galaxies reveals how much material there really is. The two don't match and there is far more matter in the intervening galaxy than we can see: dark matter.

Exactly what this dark matter is, no-one knows. We can observe phenomenon in the Universe which are explained by the concept of dark matter but we do not know what it is. If we add up all the visible matter in the Universe then it accounts for about 5% of all matter in the Universe. Dark matter is thought to account for 23%, while dark energy (remember Einstein shows us that energy and matter are one and the same thing) accounts for a whopping 72%. Unfortunately dark matter does not seem to interact with its surroundings in the same way normal matter does, so for now at least we cannot directly observe it. You might say that we are in the dark about the nature of dark matter!

Dark matter is revealed by this gravitational lens producing what is known as an Einstein Ring. (NASA)

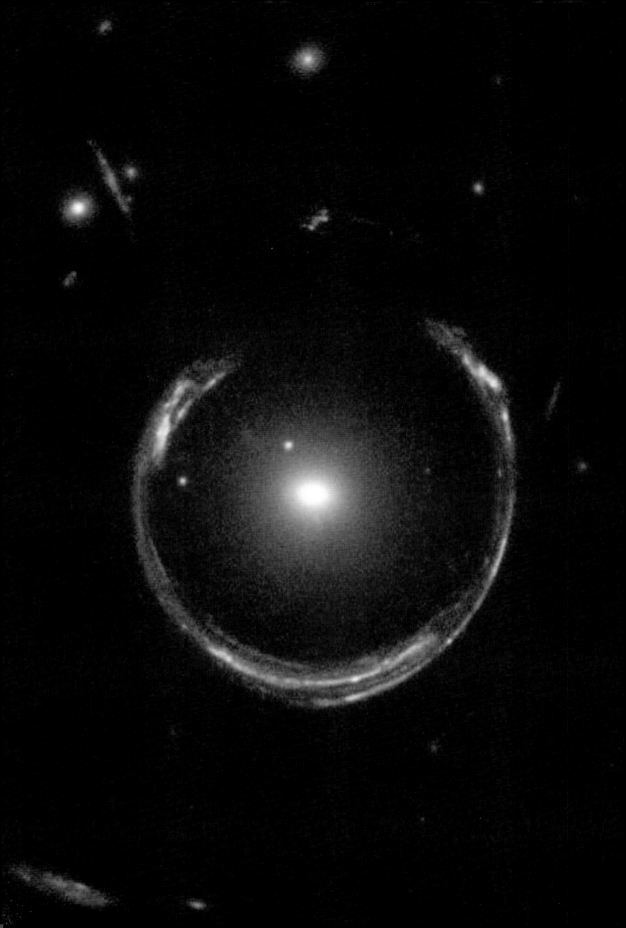

FACT 63

Studying the light from galaxies tells us how fast they are moving

If you live in a town or city then you are likely to be familiar with the sound of sirens. In particular you will more than likely have heard the sound of an emergency vehicle approaching you, heard the sound of its siren seem to change and unless you are a master criminal, will have heard the sound change again as it moves away. The movement of the vehicle causes the sound waves to get squashed up as the vehicle approaches causing the pitch to increase, and as it moves away the sound waves are stretched apart causing the pitch to go down again. Astronomers can use a similar phenomenon known as the red shift to measure the speed of galaxies as they rush through space.

If you were to analyse the sound from the emergency vehicle it would be possible to tell how fast it was travelling. Galaxies travelling through space clearly do not have sirens but what they do have is light. Astronomers use a tool known as a spectroscope to split up the incoming light from distant galaxies to reveal the detail in their spectrum.

The spectrum of stars and galaxies have dark lines superimposed upon them. The dark lines are known as absorption lines and their presence is the telltale sign of certain gasses. What is particularly useful is that the absorption lines for a particular gas exist at certain points in the spectrum irrespective of the object being studied, that is, unless they are moving. Objects that are moving away from us will have their signature absorption lines shifted towards the red end of the spectrum while those that are moving towards us will be shifted towards the blue end of the spectrum.

Unlike the earlier example of a moving emergency vehicle whose siren pitch change is caused directly by the motion of the vehicle, the red shift is related not to the movement of galaxies directly but instead to the expansion of the Universe. Imagine light waves as a spring and those light waves may have left a distant galaxy two billion years ago. As they travelled towards Earth, the very fabric of space upon which they were travelling slowly expanded. As space expanded, it stretched out the spring causing the absorption lines to shift as we see them today.

Edwin Hubble discovered the correlation between the velocity of a receding galaxy and its distance in the simple formula $D=V/H$ which tells us that we can calculate the distance to a galaxy (D) by measuring its velocity (V, from the observed red shift) and dividing it by a value known as Hubble's Constant (H). Taking this a step further we can even use this to calculate time since the Big Bang, the beginning of the Universe. If we know the distance to a number of galaxies and know how fast they are travelling then we can calculate that 13.7 billion years ago they were all in the same region of space, at the moment of the Big Bang.

The Whirlpool Galaxy and many like it have been studied spectroscopically to determine the speed they are travelling with respect to our own Galaxy.

(Mark Thompson)

FACT 64

The largest single mirror telescope has a mirror 8.2m across

Telescopes come in two principal designs: refractors constructed with lenses, and reflectors made with mirrors. Reflecting telescopes have a number of designs but the Newtonian design is most popular. The main primary mirror is situated at the bottom of the telescope tube and its purpose is to capture and reflect incoming starlight back up the tube. At the top, a flat secondary mirror is tilted at a 45 degree angle to the light to redirect it out the side of the tube where it is observed or captured by a camera.

Reflecting telescope mirrors have limitations though. The primary mirror is often a big and heavy chunk of glass. In the case of the goliath Subaru telescope on Mauna Kea it is 8.2 metres across and 20 centimetres thick, weighing in at 23 tons. That is one heavy piece of glass, making it the largest single mirror telescope. All telescope primary mirrors (including the Subaru telescope) have a very precise curve ground into them and any deformation can lead to a significant degradation in quality of image. One of the main causes of the deformation is sagging of the mirror due to its sheer weight and its changing position as the telescope moves around the sky.

Sagging of Subaru's 8.2 metre mirror is corrected constantly by an adaptive optics system. The system sits between the light coming out of the telescope and the instrument recording it, and by constantly analysing the image it calculates how much the mirror is out of shape. Attached to the back of the mirror are 36 actuators that can push and pull on the mirror to adjust its curve at the command of the adaptive optics system to correct any blur in the image.

Larger telescopes are in existence such as the Gran Telescopio Canarias but they use a slightly different style of primary mirror. Instead of one large chunk of glass, Telescopio uses 36 smaller mirrors arranged into one large primary mirror 10.4 metres across. The 36 mirrors are controlled by a system similar to the Subaru adaptive optics system to ensure the individual segments are perfect in alignment.

Neither Subaru nor Telescopio conform to the Newtonian optical style because attaching heavy equipment to the telescope at the top can cause significant balance issues. Instead they use what is known as the Nasmyth focus where the secondary mirror is placed at a point so that it aligns with one of the bearings of the telescope. This causes the light to be sent out of the telescope through the axis. Adding equipment at this point creates far fewer problems with balance.

FACT 65

Pigeon droppings were once mistakenly identified to explain the radiation from the Big Bang

In 1964, Robert Wilson and Arno Penzias were engaged in a bit of detective work at the Bell Laboratories in New Jersey. They were using a 6 metre horn-shaped radio antenna to detect radio waves bouncing off echo balloon satellites. These early communications satellites were used as passive reflectors of radio signals and would simply allow for a radio wave to be sent around the world by reflection off the giant balloons. To enable Wilson and Penzias to measure the faint echos they had to ensure that any source of interference was eliminated, and it was in this task they made a startling discovery.

Attempts were first made to remove sources of radio interference from local broadcast activity and to remove the interference from radar installations. They also removed interference from heating up of the receiver which itself was enough to cause them problems. To address this they used liquid helium to reduce the temperature of the receiver down to -269 degrees. Even with all these sources of interference removed there was still a low level hum coming from an unknown source, and this is where their detective work began.

Wilson and Penzias soon realised that no matter which direction they pointed the telescope and at what time of day, the interference remained. A pair of pigeons had been nesting in the receiver and a layer of pigeon droppings had built up inside. Suspecting the droppings could be causing the 'hum' they cleaned it all out and tried to evict the pigeons who kept returning to roost, but still the mysterious hum remained. They estimated that the interference was at a wavelength of 7.35 centimetres and that it could not have been coming from Earth, nor from the Galaxy but instead, must have been coming from deep space.

Just 60 kilometres away, a group of astrophysicists led by Robert Dicke had been hypothesising that the Big Bang must have released an incredible amount of radiation. They were about to start searching for this radiation which, due to redshift, should be visible in the microwave region of the electromagnetic spectrum. Penzias found out about this research and with Wilson, began to realise the significance of their discovery. Penzias and Wilson had detected the very microwave radiation that was the distant echo of the Big Bang and predicted by Dicke and team. They even invited Dicke over to the Bell Laboratories observatory so he could listen to the background noise for himself.

Jointly, they had discovered the cosmic background radiation, strong evidence of the Big Bang, and for their part in the discovery, Penzias and Wilson were awarded the Nobel Prize for Physics in 1978. Since those earliest days of studying the cosmic background radiation, a satellite known as the Cosmic Background Explorer, which operated between the years of 1989 to 1993, mapped the background's tiny temperature fluctuations to a much greater degree. It was replaced by the Wilkinson Microwave Anisotropy Probe in 2001 and the Planck space observatory in 2009, both of which were more sensitive and gave us an unprecedented glimpse into the early Universe.

FACT 66

Saturn's moon Mimas looks like the Death Star from Star Wars

The ringed planet Saturn is the sixth planet from the Sun and is accompanied by 82 moons. Of those, Mimas holds a special place in the heart of science fiction fans. It orbits Saturn at a distance of 185,539 kilometres, is 396 kilometres in diameter and like almost all other solid rocky moons, Mimas is covered in craters. One crater in particular has captured the imagination of the sci-fi fans. The crater, known as Herschel, was named after the moon's discoverer William Herschel, but with a diameter of 130 kilometres across it gives the moon a striking resemblance to the Death Star out of Star Wars.

Mimas, like many outer Solar System moons is composed mainly of ice with a little rock mixed in, but the high ice content has meant that tidal forces from Saturn have elongated the moon so it is rather more ellipsoid than spherical in shape. While Saturn has an effect on Mimas, Mimas has an effect on Saturn's rings. The moon is in an orbital resonance with ring particles that orbit at the distance of the A and B rings, and for every orbit of Mimas, the ring particles complete two or three orbits depending on their distance from the planet. This means Mimas exerts a tug on the ring particles on a regular basis every few orbits, causing them to move further away from the planet. The well known Cassini Division which is a gap between the A and B rings is caused by Mimas.

The surface of Mimas is pockmarked with craters but the covering is not uniform. For the most part the craters are 40 kilometres in diameter or larger, whereas particularly around the southern hemisphere there seems an absence of these larger craters and they tend to be half the size. What makes Mimas famous however is the massive Herschel impact crater.

Herschel's diameter of 130 kilometres makes it so large that it is about a third the size of Mimas itself. In areas, the crater floor is 10 kilometres deep with walls stretching up a further 5 kilometres. Most large craters have a central peak, Herschel included, which is caused by the energy of the impact. When a large rock crashes into a solid surface it causes the surface material to melt. The molten rock rushes away from the point of impact but rebounds off the solid rock surrounding the site. The rebound returns the molten rock towards the centre of the crater which then builds up the peak. The central peak in Herschel rises six kilometres above the floor of the crater.

The impact that caused the crater must have been a significant event in the geological history of Mimas and may well have come close to shattering the moon. Such was the impact event that on the far side of the moon, opposite Herschel crater, are fractures that were almost certainly caused by the impact and subsequent shockwaves.

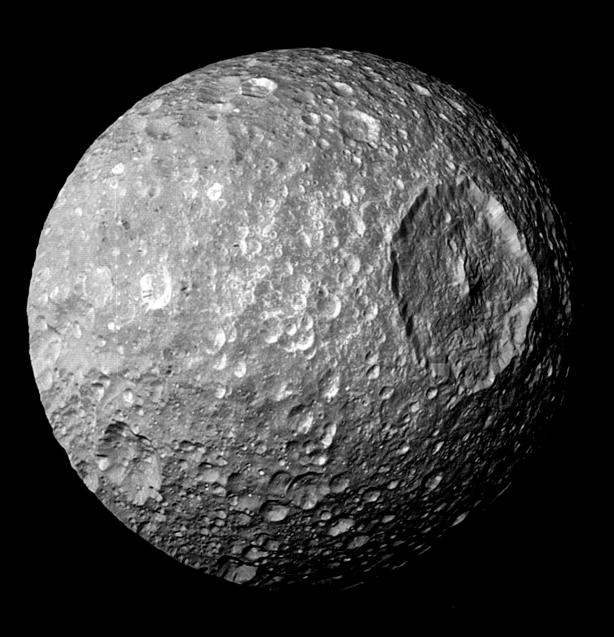

**Saturn's moon Mimas imaged by
NASAs *Dawn* spacecraft. (NASA)**

FACT 67

Neptune was first discovered by a mathematician

In August 1846 there were seven known planets in the Solar System, then by September, it had grown to eight. The planets Mercury through to Saturn were known from ancient times since they could be seen with the naked eye. Uranus was discovered almost by accident by William Herschel in 1781, who was surveying the stars when he came across it. Initially he thought it was a comet but it was soon identified as a planet. The discovery of Neptune was a little more complex.

With the discovery of Uranus, astronomers studied its movement in great detail and predicted its future position based on the known laws of planetary motion and gravitation. Future measurements of the position of Uranus were then compared to its predicted position and found to be different. Many theories were considered to account for the discrepancy, including suggestions that the laws that govern the behaviour of gravity change at great distances from the Sun! An alternative idea was that another as yet unseen planet orbited the Sun even further out than Uranus.

Astronomers, in particular French astronomer Urbain Le Verrier and British John Couch Adams turned to mathematics for the answer. Using accurate observations of the position of Uranus, they predicted the possible position and mass of the unknown planet. Neither knew that the other was working on the calculations, but using Le Verrier's predictions, Johann Galle located the planet on 23 September 1846 and he found it just 1 degree from Le Verrier's predictions and 12 degrees away from Adams'. An international debate arose from the conflicting question of who should be crowned the discoverer of Neptune but rightly, all three have that credit.

Following on from the discovery of Neptune, it and the orbit of Uranus were analysed, only to find Neptune alone could not be responsible for orbital meanderings. Calculations pointed to another planet further than Neptune. Pluto was subsequently discovered in the region of sky predicted by calculation; however, it transpired its mass was far too low. The passage of the Voyager spacecraft through the outer Solar System revised the value for the mass of Neptune, which suddenly accounted for all of the planetary wanderings. Pluto it seems just happened to be in the right place at the right time to be discovered but had very little to do with the orbital discrepancies observed in the other planets.

FACT 68

Meteorites are not hot!

Shooting stars are not stars at all, they are pieces of rock plummeting towards the Earth, but to Earthbound observers it is perfectly reasonable to think they are indeed stars that have simply gained cosmic speed and shot off across the Universe. Their rocky nature is a little less romantic and whilst they are typically represented correctly in TV shows and films, what is often misrepresented is that they are burning chunks of rock buried in the ground when they land. If you were lucky enough to have a meteorite fall at your feet then you would be able to pick it up straight away. It may be warm to the touch but would not be the smoldering piece of rock you may have been led to believe.

Finding meteorites is a tricky business, after all they do look just like the rocks in your garden. If one fell tonight and you took a look around your garden in the morning you would be very hard pushed to spot it, even if it was lying in the centre of your patio. Instead of sending expeditions to hunt for meteorites in temperate climates where meteorites would be nicely hidden among their terrestrial cousins, scientists mount expeditions to the Arctic where the hunt is somewhat easier. Finding meteorites that have landed on the white, pristine snow and ice is a little easier than hunting in the bottom of your garden. Doing this has revealed meteorites of all sorts, even meteorites that have come from Mars.

It is a common misconception that frictional forces cause meteors to glow (remember they are not meteorites until they land on the surface of Earth). People are often told to rub their hands together to make them warm and many think that this same frictional process with meteor rubbing against gas molecules in the atmosphere heats up the meteor, causing it to glow. This gives rise to the expectation of steaming, boiling lumps of rock often depicted on our screens. The reality is a little different.

Most meteoroids hit the atmosphere at speeds in excess of 100 kilometres per second. They slam into the atmosphere and compress the gas molecules ahead of them causing the gas to heat up. The gas molecules have electrons stripped off their atoms which ionises the gas, but more importantly, the hot gas surrounding the rock as it decelerates rapidly causes it to heat up. As the meteoroid heats up its surface material vaporises, and in a process known as ablation is lost into the atmosphere. It is not unusual for some meteors to look colorful as they streak across the sky; this is not an illusion, it is a real effect caused by different chemical elements in the meteor giving off light. Most meteors will be destroyed high up in the atmosphere but those that make it to the surface will have lost any heat they gained as the hot layers were vaporised, and by the time the land at your feet, they will be just warm to touch.

FACT 69

The Sun looks white in space

Let me start with a warning: never look directly at the Sun through a telescope or with the naked eye. It is dangerous and it can seriously damage your eyesight. Many of us know, however, that the Sun is yellow in colour. Ask any child to draw the Sun and invariably you will get a yellow disk (or yellow 'Sun' shaped picture), but the Sun is not actually yellow in colour, it is white.

Stars are usually categorised by their temperature and brightness or luminosity. These are combined in a diagram called the Hertszprung–Russell (HR) diagram which was first created around 1910 by Ejnar Hertzsprung and Henry Russell. The Sun's 'surface' temperature is 5,778 kelvin, and when plotted on the HR diagram, it shows as belonging to a category of stars known as yellow dwarfs, which is quite contrary to the statement earlier that the Sun is actually white.

The Sun gives off significant amounts of energy in the visible range from around 400 to 700 nanometres. Indeed this is why we humans have eyes that are most sensitive in this range. That range of wavelengths takes in all of the 'colours of the rainbow' and if you mix them together, you get white, therefore, the Sun's true colour is white. You can see this if you look at pictures of the Sun taken from space, particularly those taken by astronauts from the space station, which show a white disk.

Why then, do we still see the Sun as yellow? The atmosphere of the Earth scatters the shorter wavelengths of the spectrum such as the blue light. Subtract this from the light coming from the Sun and the previously white Sun now looks a shade of yellow. You may have seen a particular red looking Sun low down near the horizon after sunrise or just before sunset. The extra redness is the result of the shallower angle at which the sunlight is travelling through the atmosphere, scattering even more of the blue light and making the Sun appear even more red.

Some articles report the Sun as being green, but again, stars emit lots of light in all of the visible wavelengths. If a star happened to be particularly bright in the green wavelengths then the presence of blue and/or yellow would adjust the colour seen. The reality then, is that we only see the Sun as being yellow because of the effect of the atmosphere. The Hertszprung–Russell diagram can have a number of different designs but in most cases, the colour of the stars on the diagram are dependent entirely on their temperature: for example hotter stars are blue while cooler stars are red. The colours of the diagram often bear no accurate resemblance to the star's real colour but rather more the choice of the person who has created the diagram.

The effect of the atmosphere cause the Sun to look yellow to our eyes.
(Mark Thompson)

FACT 70

Voyager 1 will reach its destination in 40,000 years

5 September 1977 may have passed for many people as a fairly normal, unexciting day, but for those with an interest in space exploration it was a day when history was made. On this day the *Voyager 1* spacecraft was launched, ironically less than a month AFTER *Voyager 2*, which launched in August of that year. The mission for *Voyager 1* was to study Jupiter, Saturn, Uranus and Neptune and finally to try and locate the edge of the Solar System. It performed its task beautifully, and in 2013, it was announced that it had already left the Solar System in August the previous year. For those of us here on Earth, that was pretty much the end of the story for *Voyager 1*, but for the plucky spacecraft the real adventure had only just begun.

There is a very short window within which space probes can be launched to get them to all of the outer planets, and if not planned properly, such missions can become astronomically expensive. Instead of launching them with copious amounts of fuel to keep them going for the entire trip, mission designers make use of the gravity of the planets to not only accelerate but also change their course and send them on to their next waypoint. Gravitational slingshots like this are a great and efficient way to travel. Voyager 1 for example gained 10 km/s at Jupiter, 5 km/s at Saturn and 2 km/s at Uranus, but before these speed boosts it was not travelling fast enough to escape the Solar System. Both *Voyager 1* and *2* used gravitational slingshots effectively but if the launch window was missed they would have had to wait a further 175 years for another so called 'Grand Tour Alignment'.

Voyager 1 has exited the Solar System and is travelling at a speed of 62,140 km/h. It is now at a distance from Earth of 21 billion kilometres but it has a long way to go before its next scheduled encounter. In 40,000 years, yes FORTY THOUSAND YEARS, *Voyager 1* will fly within 1.7 light years of a star known as AC +79 3888 or Gliese 445 in the constellation Camelopardalis. The star is 17.5 light years from Earth so any signal sent to the spacecraft will take 17.5 years to get there. Sending signals to *Voyager* though will be a waste of time because the probe's power supply will run out in less than 10 years, so none of the equipment will be functioning.

Any alien civilisation that happens to pickup our interstellar messenger will still receive a coded message from Earth, assuming they can work out how to interpret it. Attached to *Voyager* is a golden record upon which are visual representations of how to play the record and where it came from. Any alien that works out how to play the record will be treated to the sounds and greetings from Earth in 54 languages. There are 117 images and a selection of sounds that include thunder storms, volcanoes, rocket launches, aircraft sounds and animal noises. It is amazing to think aliens inhabiting distant worlds may one day listen to sounds from Earth but it is perhaps more amazing that our own future space travellers may well beat Voyager 1 to its destination. Rocket technology has developed at an incredible rate in just a few decades; imagine how different it will be in 40,000 years!

FACT 71

We are sending signals into outer space and aliens could be listening

Knowing that we are not alone in the Universe would be an incredible discovery. The social, psychological, and not to mention religious, implications could be life changing. We are searching for alien civilisations by hunting for exoplanets, listening for radio signals and even scouring our own Solar System for signs that we are not alone, but it is just possible that aliens already know we are here, without us knowing about them!

Efforts have already been made to announce our presence to the cosmos. For example, in 1974 the Arecibo radio dish transmitted a signal towards the globular cluster in Hercules known as M13. Interstellar communication is largely a waiting game though, because M13 is 22,180 light years away, so this means any civilisation living on a planet orbiting one of the stars will not detect the signal for another 22,136 years. If they decide to reply straight away then we will have to wait 44,316 years from today! The message was sent more to demonstrate human achievement and capability more than it was a serious attempt to make contact but the signal contained some useful information to anyone that could decode it. Among other things, it carried details of where the message came from, the inhabitants that sent it, along with various bits of scientific information to show we are an 'intelligent' civilisation.

Whether the signal from the Arecibo dish gets picked up by anyone is difficult for us to predict. After all, the transmission lasted for only three minutes so anyone out there in the right direction needed to be listening at the right time and at the right frequency to pick up the message. Other signals have been sent that are far more likely to announce our presence to the Universe – and we have been sending them every day since 1936!

In 1936 the Olympic Games were held in Berlin in Germany and the opening ceremony was televised and beamed around the world. The games were the first TV transmission beamed at a frequency that could escape through the atmosphere. In theory, any civilisation within about 80 light years of Earth might be able to pickup that and all other TV shows that heave been beamed around the world since. Or can they?

For a signal to be readily distinguishable from the general noise of space it needs to be strong and clear otherwise it could be easily overlooked. Even a strong signal that has been purposely sent out into deep space it will get weaker the further it travels, but for TV signals that have 'leaked' into space, it is not likely, not impossible but not likely, that they are going to be picked up by civilisations that might be looking at distances greater than just a few light years. The chances have reduced further in recent years with the advent of digital TV transmission technology. The reality is that any aliens out there on the lookout for others are more likely to detect our presence through careful analysis of the light from our Solar System than they are from tuning in to our soap operas or reality TV.

Image of globular cluster M13, the destination of
the Arecibo radio transmission. Image captured
by the Hubble Space Telescope.
(NASA, ESA and the Hubble Heritage Team [STScI/AURA])

FACT 72

Mercury is shrinking!

Mercury is the nearest planet to the Sun, and since the demotion of Pluto to the status of dwarf planet, it is now also the smallest. It rattles around the Sun in just 88 days and measures a tiny 4,879 kilometres in diameter. Compare this to the diameter of the Moon at 3,474 kilometres and you realise just how small it is. It might come as no surprise then that there is strong geological evidence that Mercury is shrinking.

Cliffs, fractures and folds are common in the Solar System and a common effect of plate tectonics. Even on Earth we can see evidence of the movement of plates causing gigantic mountain ranges to be forced out of the ground. Spacecraft have visited and studied Mercury at close quarters and have identified many such features, known as rupes. They can be the result of tectonic activity or the result of a planetary crust cooling and shrinking over millions of years in just the same way that an apple skin wrinkles up as the apple dries and shrinks.

The rupes on Mercury suggested cooling was the cause, however there was no other evidence that the planet had actually shrunk. The Messenger spacecraft changed that view during its studies of Mercury between 2011 and 2015. Its four-year mission at Mercury allowed scientists to closely study and profile the terrain of the innermost planet. Messenger discovered many more new features like scarps and the wonderfully named wrinkle ridges.

Earth based observational data has helped to determine Mercury's volume and from careful analysis of spacecraft telemetry revealing how they were tugged at by Mercury during the many orbits enabled Mercury's mass to be calculated. Armed with these two crucial bits of information it is possible to calculate the density of Mercury. The density of Mercury is only a little less than that of Earth, but Earth's density can be attributed to gravity compressing our iron core. Mercury is much smaller so does not have sufficient gravity to compress its core in the same way. Instead it is thought that Mercury must have a large core composed of heavy elements to explain its density. Current estimates suggest Earth's core is 17% the volume of the planet whereas Mercury's core is 42%.

Taking into account a global view of the matter distribution and internal structure of Mercury, its crust is estimated to be 250 kilometres thick. Consideration of the various geological features observed over the planet's surface leads to only one conclusion, that Mercury is shrinking, and since its formation 4.5 billion years ago, its thought to have steadily shrunk by 14 kilometres. That Mercury is shrinking is not necessarily surprising news because the cooling of any object generally leads to a reduction in size. Scientists have even debated whether Earth might be shrinking too but the jury is well and truly out on that one.

Beagle Rupes, the tallest and longest scarp on Mercury, caused by the planet's cooling interior. (NASA)

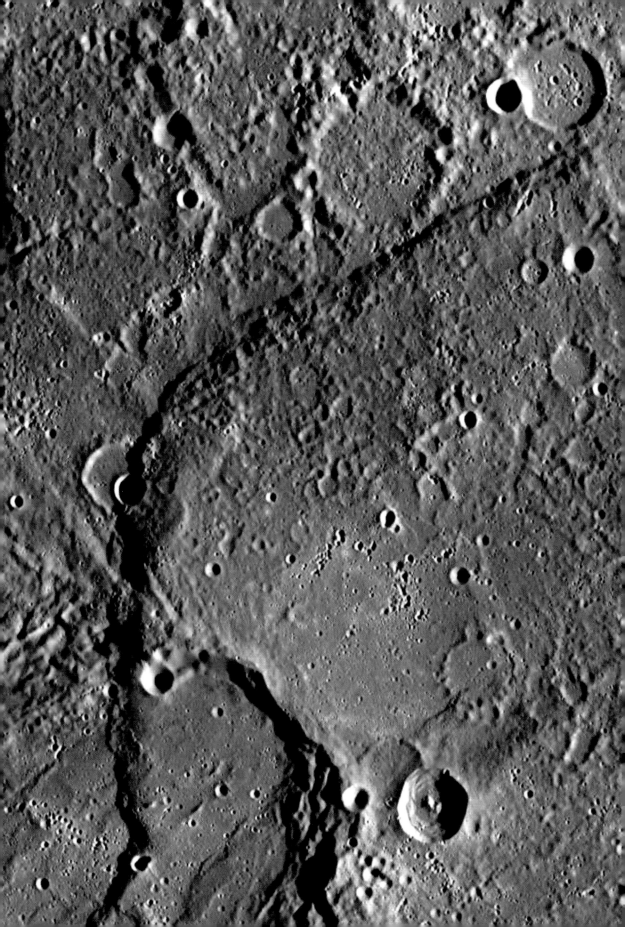

FACT 73

The Hubble Space Telescope was launched with the wrong mirror

The Hubble Space Telescope (HST) was launched into Earth orbit on 24 April 1990. The concept of putting a telescope up into space was first discussed in 1946, 44 years before Hubble was launched. The space telescope, named after astronomer Edwin Hubble, was put up into space for numerous reasons: to escape from the constraints of the weather, to get above the absorbing/distorting effects of the atmosphere and to escape light pollution. It was a great but expensive concept that history has shown now to be hugely successful. Over the years, the HST has given us some exquisite images of the Universe, but the first image was less than impressive.

The telescope is approximately the size of a double decker bus but at its core is the 2.4 metre mirror. Light from distant objects falls down the tube, strikes the parabolic mirror, reflects back up the tube, strikes a second mirror which is convex that directs the light back down the tube, through a hole in the primary mirror and into the various instruments sat behind. The telescope was designed and built to a high degree of precision and cost $1.5 billion dollars. Unfortunately, whilst the primary mirror was built to a very high degree of accuracy and was polished to the right level of smoothness, the shape of the curve was wrong! To the dismay of the engineers, when the first images were received from HST, they were blurry. Suspecting focus problems, they tried to focus the telescope but could not: it seemed the telescope mirror suffered with a problem known as spherical aberration.

To get a sharp image through a telescope, all incoming rays of light need to be brought to a focus at exactly the same point. If their various points of focus are at slightly different places then it is impossible to get a sharp focus and blurry images will be the best that can be achieved. To overcome this problem with reflecting telescopes like Hubble, they have a slightly deeper curve in the centre than at the edges, a curve known as parabolic. Hubble's mirror needed a parabolic curve but analyses of the images revealed that the curve was wrong and was out by one millionth of a metre or 1/50th the width of a human hair! Investigations revealed some of the optical testing equipment used to asses the mirror's curve was incorrectly calibrated.

Replacing the telescope mirror was simply impractical and expensive so NASA launched a servicing mission in 1993. The objective of the mission was to install COSTAR, the Corrective Optical Space Telescope Axial Replacement system. It contained small mirrors that were mounted on robotic arms that were used to correct the incoming light beams before they entered into the cameras and instruments. It worked beautifully and Hubble could finally see. All instruments that have been installed since have been designed to correct the optical defect, so COSTAR was removed in 2009.

The galling and cruel twist in this story is that a second, backup mirror had been commissioned by NASA. It never formed part of the mission but testing has shown that not only was it made to the correct smoothness but it was also made with the correct curve. If it had been used, none of the trouble with Hubble would have been experienced.

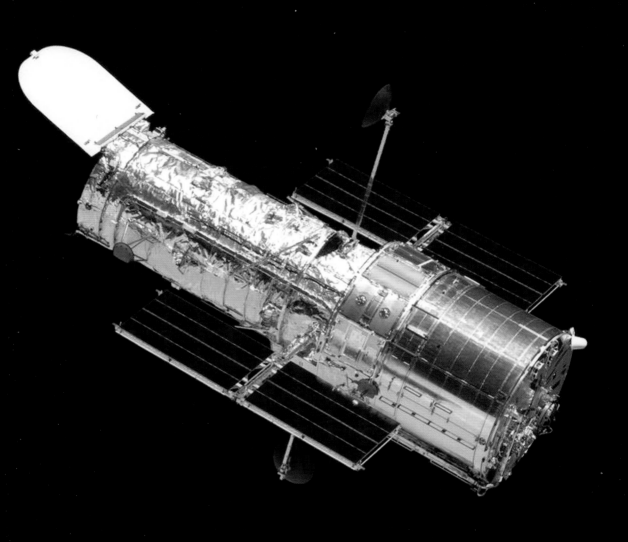

Hubble Space Telescope. (NASA)

FACT 74

A veritable menagerie of animals have been sent into space

A lot of people will have heard the name Yuri Gagarin, the first human in space, who flew on 12 April 1961. Before humans sent one of their own up into orbit, animals paved the way for human spaceflight. Ask anyone what animals have been into space and you will probably get an answer that includes dogs, monkeys and perhaps mice, yet the list is considerably longer.

Laika the dog is probably the most well known animal that was sent into Earth orbit, on 3 November 1957, just four years prior to Gagarin's historic flight. Launched into space on board the Russian *Sputnik 2* spacecraft, Laika sadly died in space, as (horribly) planned, because the technology to return her safely to Earth had not been developed. Instead of a humane death, she died from overheating just hours into the flight. Laika was not the first animal into space though, nor was she the first animal to go where humans had not gone before. Even as far back as 1783, the Montgolfier brothers sent a duck, rooster and sheep up in a hot air balloon.

The first creatures into space were fruit flies on board an American V2 rocket in February 1947. They launched from New Mexico with a mission objective to study the effects of radiation at high altitudes. They reached an altitude of just over 100 kilometres in a little over three minutes before being returned safely to Earth by parachute. After the success of the fruit flies, it was the turn of the primates.

Just two years after the fruit flies' successful sub-orbital flight, a rhesus monkey named Albert was sent up on a V2 rocket. Albert died on ascent having reached an altitude of around 40 kilometres, so he was followed by Albert II who reached 134 kilometres. He died on returning to Earth during an impact following parachute failure. Mice were then launched on board American V2 rockets and dogs were launched aboard the Russian R-1 rockets. This brings us up to date with Laika who became the first creature to achieve orbit.

The first animal space crew was launched on 28 May 1959 and was made up of another rhesus monkey named Abe and a squirrel monkey named Baker. Together they reached an altitude of 579 kilometres and experienced 38 times the normal pull of gravity (meaning they would have felt 38 times heavier than usual) and experienced 9 minutes of weightlessness. The total flight lasted for 16 minutes and both monkeys survived their ordeal, although died later from various other conditions.

Next came a chimpanzee called Ham. He was launched on board a Mercury capsule atop a Redstone rocket in January 1961. He had been trained to pull levers to receive rewards and avoid painful electric shocks. The flight was designed to show that tasks could be performed successfully in space. Just three months after Ham's flight, the first American astronaut, Alan Shepard, went into space on a sub-orbital flight.

Rats, frogs, guinea pigs, cats, parasitic wasps, beetles and stick insects soon followed but the real space explorers were two tortoises, some wine flies and meal worms that were sent on a circumlunar voyage around the Moon and returned unharmed except for a little weight loss.

FACT 75

The movement of the Earth allows astronomers to measure distances in space

Measuring distances in space is a tricky business. If you want to know the distance to something here on Earth then you could measure it with a tape measure or for longer distances use the distance readout in your car. The vehicle relies on measuring the number of rotations of a circle of known circumference which can be translated into distance. It is not quite that easy when it comes to working out distances in space.

Most people are familiar with the annual movement of the Earth around the Sun as it travels the 940 million kilometres around its orbit. The average distance between the Earth and Sun is 150 million kilometres so the separation between any two points opposite to each other on Earth's orbit is around 300 million kilometres. Astronomers use this as a base upon which to determine the distance to nearby stars. You can try this for yourself. Extend one of your arms out in front of you and raise a finger pointing it up towards the sky. Now shut one eye and notice the position of the finger against the background you can see. Now open that eye and close the other, you will see the finger appears to have shifted position without even moving. If you really wanted to you could measure the distance between your eyes, measure the angular shift of your finger and calculate the length of your arm using simple high school trigonometry.

In the example above, using a tape meaure would be much easier to work out how long your arm is but we cannot do that for stars. Instead we use the extremes of the Earth in its orbit six months apart to take measurements of the position of stars in the sky. Knowing the distance between the two points and being able to measure the shift or 'parallax' of the star means we can determine its distance.

This is not a recent thing. Hipparchus used a similar technique to calculate the distance to the Moon in 189bc. A total solar eclipse was visible from Turkey but in Alexandria in Egypt the Moon only covered 4/5ths of the Sun. Knowing the distance between the two was 965 kilometres and measuring that the edge of the Moon was about 1/10th of a degree further over from Turkey he calculated that the Moon was 563,300 kilometres away. His answer was quite wrong, about twice as far as later measurements, but given that this was done over 2,000 years ago it is pretty impressive.

The first to measure the distance to a star using this stellar parallax was F.W. Bessel, who measured the distance to 61 Cygni in 1838. Its tiny parallax shift of 0.28 arc seconds (there are 60 arc seconds in one arc minute and 60 arc minutes in one degree) meant it had a distance of 11.6 light years, which is impressively close to today's measurement of 11.4 light years. Among the most distant stars that have had their distance determined by stellar parallax are stars studied by the Hipparchos telescope, which calculated distances of stars out to a distance of around 1,600 light years.

FACT 76

Stars do not live forever

Look up at the night sky and you would be forgiven for thinking that the stars will be there forever. It is not an unreasonable assumption given that the stars you see as a child are the same stars you will see once you near the end of your life. Moreover, the stars you see today will be the stars that your grandparents, even great grandparents will have seen. The reality however is that our lives are short in comparison, maybe 100 years if lucky, yet the average star will live for maybe 10 billion years.

The death of a star is closely linked to the mass of the star. Take our own local star the Sun, a fairly average star that is fusing hydrogen into helium in its core. The process of fusion generates an outward pushing force known as the thermonuclear pressure. For most of the Sun's life and indeed any star's life, when this force balances the force of gravity trying to collapse the star then it has reached a period of stability. In the case of the Sun these forces are now in a state of equilibrium and the Sun remains stable.

Eventually the core will run out of hydrogen and the thermonuclear pressure will diminish, gravity starts to win and the core contracts. This contraction causes an increase in core temperature leading to an expansion and cooling of the outer layers, turning it into a red giant. Slightly more massive stars may now start fusing helium into carbon and oxygen in their core but at this stage of the star's life, much of its outer layers can be lost into space, crafting the beautiful planetary nebulae. All of the star's outer layers will be lost to space and the carbon core of the star left exposed will slowly cool and fade.

Stars that are eight times more massive than the Sun will experience a rather more dramatic death. The carbon core will continue to contract, raising the temperature and pressure again so that carbon can fuse to neon. The process will continue, with core contraction leading to new phases of nuclear fusion that generate successively heavier elements until the core is made of iron. It is not possible to fuse iron so the thermonuclear pressure ceases and gravity finally wins, leading to further core collapse.

The next step in the process depends yet again, on mass. If the core has a mass less than three times the mass of the Sun then the force of gravity is not strong enough to overcome the strength of neutrons (imagine trying to crush a brick with your hands, you are not strong enough and the bricks survive, but put them in a gigantic press and the bricks will get crushed). The remaining core is effectively one massive neutron that we call a neutron star. The sudden halt in contraction causes a shockwave to rebound outward, a supernova explosion. It is this process that scatters the heavy elements throughout the Universe and which ultimately leads to the formation of complex organisms like you and I. For stars with a core mass greater than three times that of the Sun the strength of the neutrons is not enough to halt the collapse, so gravity continues unchecked, collapsing the core into a single point, a singularity. These collapsed cores are infinitely dense and are known as black holes.

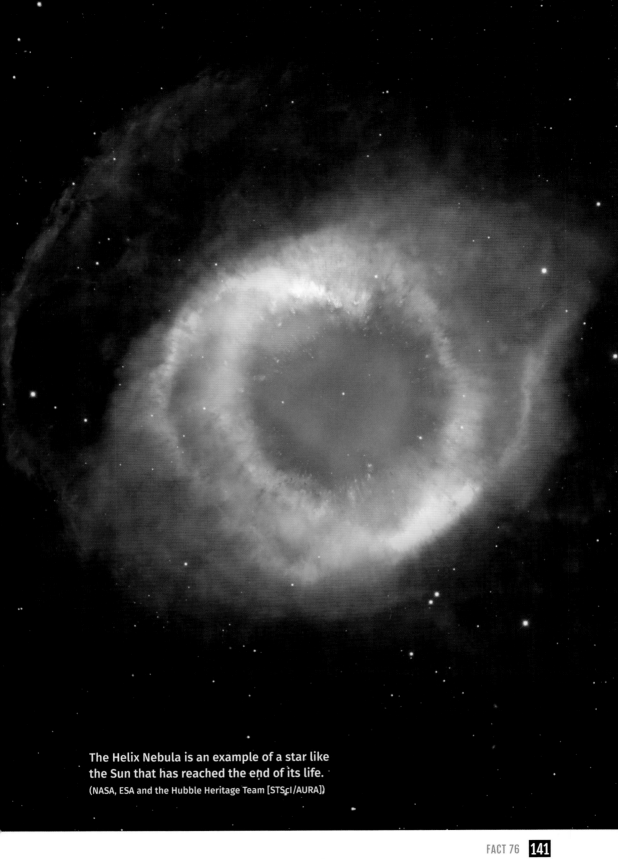

The Helix Nebula is an example of a star like
the Sun that has reached the end of its life.
(NASA, ESA and the Hubble Heritage Team [STScI/AURA])

The different shapes of galaxies relate to their evolutionary stage

If you have ever taken a flick through a book with pictures of the Universe you will undoubtedly have seen images of different types of galaxy. Our own Milky Way galaxy is thought to be a type known as a barred spiral galaxy, there are the more simpler form of spiral galaxy and also elliptical galaxies.

For many years, astronomers thought these were different types of galaxy but with clearly defined characteristics. Spiral galaxies for example can be visualised by imagining two fried eggs stuck back to back with the yokes of the egg representing the nucleus of the galaxy and the egg white the plane within which we find the spiral arms. From above (or below) a spiral galaxy looks like a giant catherine wheel on bonfire night. Barred spiral galaxies like the Milky Way are the same but with a very distinctive bar shape protruding out of the nucleus and from which, the spiral arms radiate. Elliptical galaxies as their name suggests are shaped like a rugby ball but other than that, they have no other defining shape and certainly no spiral arms.

In 1926 Edwin Hubble devised a diagram which became known as the Hubble Tuning Fork Diagram due to its visual similarity with a musical tuning fork. The diagram showed how each of the three classes of galaxy had quite a bit of variability: for example spiral galaxies can have their spiral arms very tightly wound or more loose, and elliptical galaxies can be rugby ball shaped or more spherical.

Observations of galaxies over the years started to reveal some interesting features that hint at the connection between them all. Spiral galaxies were found to be brimming with active regions of star birth whilst elliptical galaxies seemed full of aging stars with very few, if any, stellar nursery. NGC3359 is a galaxy which at first glance falls into the barred spiral category but it seems that the bar has only recently (in galactic terms) formed. NGC1300 is another barred spiral galaxy that still seems to hint at a spiral structure at the core, suggesting an evolutionary path between the two.

It is now believed that many galaxies exhibit properties similar to spiral galaxies in their youth but as they evolve, they develop a bar. Barred spiral galaxies seem to have their fair share of stellar formation and older stars so these seem to represent middle-aged galaxies. Further galactic evolution seems to develop barred spiral galaxies into elliptical galaxies which nicely explains why stellar formation seems to be largely absent in these.

A beautiful pair of interacting galaxies called ARP273.
(NASA, ESA and the Hubble Heritage Team [STScI/AURA])

FACT 78

The rotation of the Sun is not uniform

Consider the spin of the Earth: it takes 23 hours, 56 minutes and 4 seconds to complete one rotation around its axis. We approximate this to give us the 24 hours in one day. If you stood just one metre away from the axis of rotation, at either the north or south pole, then you would take the same time to complete one rotation as someone stood on the equator. Clearly the equator must be travelling much faster than the polar region to be able to complete one revolution in the same time. If it was not, then our world maps would constantly need updating as the relative positions of countries kept moving around!

The scenario described above is how solid objects rotate, with one country remaining next to another country as it rotates. For objects like the Sun that are not solid then regions that are next to each other will not remain next to each other through subsequent rotations. This is known as differential rotation and in the case of the Sun, the polar region completes one rotation in 35 days but the equator completes a revolution in 25 days.

It is often said that the Sun is a large ball of gas measuring 1.4 million kilometres in diameter. Gas is defined as 'an air-like fluid substance which expands freely to fill any space available regardless of its quantity' but the Sun and other stars are made of a particular type of gas known as a plasma, a gas which is electrically charged. Electrically charged gas and differential rotation make for some interesting effects, particularly when combined with a body that has a magnetic field.

The Sun, like Earth, has one such magnetic field and it runs from north pole to south pole around and through the Sun. Now imagine the Sun frozen in time with neat magnetic field lines running from north pole to south running perpendicular through the solar equator. Unfreeze time and the Sun starts rotating with the moving electrically charged gas dragging the magnetic field lines along with it. The faster rotation rate of the plasma at the equator (25 days) causes the field lines to get dragged ahead of the lines in the polar regions where the plasma is moving at a slower rate.

As time progresses the field lines get wound up tighter and tighter and are subjected to more and more stress. Some of them snap or burst through the visible surface of the Sun, leading to the formation of sunspots. The complex maze of field lines build up over the periodic 11-year solar cycle until the lines cannot take any more stress and snap back to their original starting position, before the cycle starts again.

Observationally we can see this process in action. At the start of the 11-year solar cycle, sunspots are at a minimum. They start to appear around mid-latitudes and slowly migrate towards the equator as the field lines get more wound up. We can even observe regions where field lines have burst through the visible solar surface dragging solar plasma with them. Eventually the process reaches its end and the sunspots fade, marking the start of a new solar cycle.

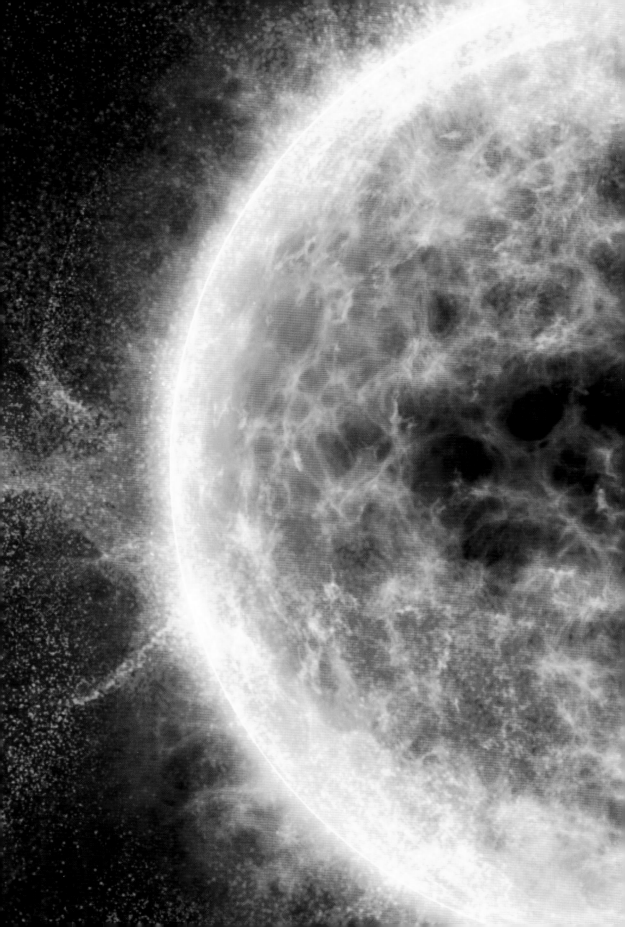

FACT 79

Some variable stars can tell us how far away they are

Stars are well known for being twinkly, but the twinkle of a star that you can see as it flickers against the blackness of space is the effect of the Earth's atmosphere. Some stars do vary in the amount of light we receive from them. For some it is because of an unseen companion, and these are known as eclipsing binaries. Others vary in the amount of light they emit. This latter type are known as variable stars and there are numerous different types. One type of variable star is known as the cepheid variable and not only is it an interesting type of star to study but it is useful in gauging distances in space.

Cepheid variable stars pulsate, increasing in size and luminosity over a very regular period. Helium gas is thought to be a key player in the pulsating mechanism. If you heat helium then it will give one of its electrons too much energy, be lost to the surroundings, become ionised and become a little opaque. Heat it further and the second electron will be lost, the helium will be said to be doubly ionised and it will be even more opaque. When a cepheid variable is at its faintest, the helium in the outer layers has been double ionised by heat from the star and is therefore opaque, letting less light or radiation through. Due to the increased temperature, the outer layers expand and cool, causing the helium to become less ionised and less opaque, letting more radiation through again and the star brightens. The expansion stops with the escape of the radiation, gravity momentarily overcomes the star, shrinking it back to its original state, and so the process continues.

This constant and reliable pulsing of the star is closely linked to the amount of radiation that the star emits. If a star is studied over a period of time then it is possible to identify a cepheid variable by the way in which the light output changes over time. This can be done regardless of the distance to the star as long as its light can be seen. This is where the cepheid variable comes in useful for determining distances, and because they are 100,000 times more luminous than the Sun they can be seen over vast distances, even in nearby galaxies.

Identification of such stars in a nearby galaxy can be done by studying how the star's light output varies over time. Their actual brightness is very closely related to how they vary so identifying one from its light curve will allow us to infer the luminosity. We can then compare the brightness of the star in the sky to how bright it really is to calculate its distance. Using cepheid variable stars is an easy and reliable way to measure distances in space. Its limit is defined only by how far away they can be detected. To date, the most distant cepheid has been found to lie 108 million light years away.

**RS Puppis is a Cepheid Variable class of star
in the constellation Puppis.**
(NASA, ESA and the Hubble Heritage Team [STScI/AURA])

FACT 80

Spacecraft docking in space is a tricky manoeuvre

It is fair to say that spaceflight itself is challenging, but trying to make two spacecraft dock whilst in orbit is even more tricky. Whilst it is perhaps one of the hardest manual manoeuvres to perfect it is a quite common. Spacecraft docking techniques were first developed by NASA, whose original plan was to manually dock the crewed *Gemini 6* craft with an un-crewed Agena vehicle. Unfortunately the Agena vehicle exploded on launch, but later attempts were successful. From Apollo missions to the Moon to crew changes on the International Space Station, docking is a routine aspect to life in space, but there are challenging orbital mechanics behind it.

Imagine seeing a friend walking down the street and you want to catch them up. What do you do? You walk a bit faster or run to catch them and before long you are chatting away to them, maybe a bit out of breath. You might think that docking spacecraft is as simple as this: if you are the commander of the spacecraft following then you accelerate to catch up. Unfortunately it is not as simple as that. Anything in orbit, be it a spaceship or a moon, will be travelling at a certain speed to maintain that orbit. If it slows down then its orbit will slowly decay and it will eventually fall back to Earth. Similarly if it accelerates then it will slowly drift into a higher and higher orbit. You may now be able to see the difficult. If you are in the same orbit and accelerate your spacecraft you will certainly catch up but you will drift into a higher orbit and shoot 'over the top'. Docking spacecraft requires starting the manoeuvre when the following craft is behind and lower than the vehicle to be docked with and a careful juggling of speed against altitude.

The process of docking spacecraft involves two distinct phases. First is the soft docking. In a soft dock, contact is made between the two modules and the docking mechanisms on both latch together so they are now connected and will not drift apart. Once the soft dock is secured the hard dock is initiated. This creates an airtight seal between the two spacecraft so that hatches can be opened and crew or cargo can be safely transferred between them both.

Perhaps the most incredible docking manoeuvre of all time took place in 1985 when the *Salyut 7* station ran out of power and ceased operation. Russian commander Vladimir Dzhanibekov and engineer Viktor Savinykh took *Soyuz T-13* to effect a repair. Typically, Russian spacecraft used an automated docking system, but the power failure meant the station was not transmitting any telemetry, meaning they had no information about speed/position so manual docking was the only option. In one of the most complex manual docking procedures they measured distance using hand held laser rangefinders, matched the rotation of the station and eventually established a soft dock. They confirmed the station had lost power, established the atmospheric conditions were acceptable and, wearing winter clothing, entered the cold, dead station. After a week making repairs the station was brought online again.

Without the ability to dock spacecraft while in orbit, human exploration in space would be significantly more challenging.

The *Apollo 10* command module about to dock with the lunar module. (NASA)

FACT 81

The Universe is 92 billion light years across

Any debate about the size of the Universe should first define what we mean by 'the Universe'. In considering its size we can only define what we can see and we call that the observable Universe. This fact should therefore be retitled to read 'The observable Universe is 92 billion light years across'! It is defined chiefly by the speed of light and the age of the Universe since we can only see things where there has been sufficient time since the Universe formed for their light to reach us. It does however keep getting bigger, as the older the Universe, the further light can travel and the further we can see. In this definition, the observable Universe is currently 92 billion light years across.

The key to reaching this conclusion is to observe the light, or more accurately the electromagnetic radiation, that has travelled for the longest time: the cosmic background radiation (CBR) which is the echo of the Big Bang itself. The CBR covers the entire sky and in 2013 the most accurate map of it was released from observations from the Planck Space Observatory in microwave and infra-red frequencies. Analyses of the radiation in the CBR revealed that the Universe is 13.8 billion years old, which means we can study the Universe at a distance of 13.8 billion light years or 1.3×10^{23} kilometres away in any direction.

This makes the observable Universe 27.6 billion light years in diameter, but the Universe has been expanding over the entire course of its history. If we make the assumption that the expansion has been constant for the majority of this time then it tells us that the same spot that lay 13.8 billion light years away now lies 46.1 billion light years away, making the Universe 92.2 billion light years in diameter.

It may even be larger than this because we are basing this on what we can see. It also suggests that we are at the centre of the Universe and whilst that is a typically human consideration that everything revolves around us, history has shown this never to be the case. Consider the analogy of a ship on the Atlantic Ocean with no sign of land in any direction. The visible horizon may be only five kilometres away but that does not mean the ocean is only 10 kilometres across nor does it mean you are at the centre of it. In reality it is just under 5,000 kilometres across.

From our vantage point the observable Universe is 92.2 billion light years in diameter but we should take into account the shape of the Universe too. Recent studies suggest the Universe is flat and if this is the case then the true extent of the Universe beyond what we can see may well be infinite.

FACT 82

In 1977 astronomers thought they had detected an alien radio signal

On 15 August 1977, the Big Bear observatory, operated by Ohio State University, detected a signal that COULD have forever changed the world in which we live. The signal had gone unnoticed for a couple of days when Jerry Ehman saw the computer printout and was so shocked at what he saw that he circled the data and scribbled 'Wow!' next to it. He and his colleagues thought they had discovered a signal from an alien intelligence.

The telescope had been pointing towards the constellation Sagittarius when it recorded the signal. Being a radio telescope the readout simply showed strength of signal. The alphanumeric sequence on the readout represents the strength of the signal averaged over previous minutes. The strength ranged from 0 to 36 with a code to make analysis easier, where a space on the readout represents the weakest signal: 0–1. A signal with a strength 1–9 is represented by their corresponding number and a signal strength of 10 or more is represented by letters. Signals of 10 or 11 are shown as 'A', 12 to 13 is represented as 'B' and the highest signal, over 30, is shown as 'U'.

The signal that caused such excitement bore the sequence 6EQUJ5, and notice it contained a 'U' which is a signal strength 30 times greater than background noise. It is important to note here that the sequence of letters bear no resemblance to any secret coded message, but simply represents a burst of radio energy that lasted for 72 seconds.

It is difficult to identify exactly where the signal came from because the Big Bear observatory consists of two horn-shaped receivers and both point at slightly different regions of the sky. It is not even possible to determine which receiver picked up the signal. Several attempts have been made to pickup the signal again with the Big Bear telescope but these have all been unsuccessful. Further searching was undertaken as late as 1996 using the Very Large Array radio telescope interferometer, a system that is more sensitive than Big Bear, but it too was unable to find anything, nor were studies using the Mount Pleasant Radio Observatory 26 metre dish.

We may never know where the signal came from. One theory suggested its origin might have been terrestrial and that it bounced off a satellite or piece of space debris in orbit around the Earth. However the frequency of the signal was in a bandwidth that is restricted, and transmissions at this frequency on Earth are forbidden because astronomers use it for peering deep into space. It is therefore very unlikely that this came from Earth. It seems then that a transmission from space is the likely cause of the signal but whether the origin is alien is something we may never know.

Gamma ray bursts are among the most powerful explosions in the Universe

In the late 1960s the US Vela satellites were operational and searching for signs of secret space-based nuclear tests. They were on the lookout for bursts of gamma radiation, the telltale sign of a nuclear explosion. In July of 1967, the Vela 3 and Vela 4 satellites detected one such burst of gamma radiation but it was not like any previously seen flash from manmade nuclear bombs. The flash was ignored for some time but more sensitive Vela satellites were launched that detected even more unidentified bursts of gamma rays. The analysis of the bursts' detection from different satellites revealed their approximate position in the sky, ruling out anything created by humans, ruling out anything Earth-based in origin and ruling out anything based in our Solar System. They were coming from deep space!

We had to wait until April 1991 and the launch of the Compton Gamma Ray Observatory with its plethora of instruments to hunt down these strange and elusive bursts to start to unravel their innermost secrets. On board Compton was the Burst and Transient Source Explorer, which was incredibly sensitive and identified and revealed the location of a number of gamma ray sources. The results revealed that the gamma ray bursts (GRBs as they soon became known) were coming from all directions. This suggests, that although it is not fully conclusive, that they come from outside the Milky Way, otherwise their distribution is more likely to be focussed around its structure.

The bursts can be very short lived (even just a few milliseconds) to something lasting an hour or more. Following the burst there is often an afterglow at longer wavelengths: UV, visible, infra-red or radio, but none have been positively matched with any other object. Careful study of the region of the sky from which the GRBs have been observed reveals nothing in visible or other wavelengths. This suggests that the bursts are perhaps the brightest events in the Universe occurring at distances so vast that we simply cannot see the objects they are coming from.

Some recent observations have drawn a connection between GRBs and hypernova explosions. On 29 March 2003 a burst of gamma rays was detected, named GRB030329, but a few days later, spectral analysis of the event started to reveal new spectral features consistent with a hypernova explosion. The collapse of the core of a massive star leads to the formation of a black hole, and at the instance of collapse, it is thought that jets of material shoot out of the core, slamming into the outer layers of the star, generating high temperatures and releasing gamma rays. The jets continue outward into less dense material resulting in the longer wavelengths as observed in the afterglow. This model fits the observation of some GRBs, but others, which seem to be at far greater distances, still hold mysteries that we are yet to uncover.

Gamma Ray Burst GRB 151027B captured by NASA's *Swift* spacecraft. (NASA)

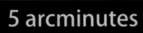

5 arcminutes

FACT 84

The constellations of today will not be recognisable in a hundred thousand years

The patterns in the stars which we call the constellations are a human construct. Throughout the centuries, different cultures have seen patterns in the stars which have in many cases reflected their mythological stories. The International Astronomical Union recognises 88 constellations, and over half of them came from the ancient Greeks. In depicting them we tend to visualise them in two dimensions but in reality the stars that make them are at different distances. For example, the three belt stars in Orion line up beautifully in our skies yet the left star Alnitak is 1,262 light years away, the central star Alnilam is 1,344 light years and Mintaka the right star is 916 light years away. Move to a different part of space and they would look very different.

We tend to assume the constellations are never changing since, other than different ones being visible at different times of year, they seem to maintain their relative positions. You could even come back in 100 years or maybe 1,000 and they will look the same But come back in 100,000 years and a very different, almost alien sky would greet you.

Stars move through space; even our Sun is hurtling around the Galaxy at 828,000 kilometres per hour, but consider this. Next time you are sat in a car as a passenger, look out of the side window and you may get to see nearby and distant trees. You are travelling past both at around 100 kilometres per hour but because they are at different distances their angular speed seems to be quite different. The trees nearby seem to whizz past in a bit of a blur while the more distant ones barely seem to move. It is the same with the stars, they are travelling fast, very fast in comparison to the car analogy but because their distances are somewhat greater, they take ages to appear to move.

Barnard's Star is the fastest moving star in our night sky, but even then, to travel the distance equivalent to the width of the full Moon takes it 180 years! The speed of a star across the sky is a function of its true speed, its distance and the direction it is heading. Barnard's Star is 5.9 light years away so its proximity is one reason we see it moving fast, but more distant stars, such as those in Orion's belt, will take thousands of years to traverse the same distance.

One of the most well known star asterisms (a pattern of stars smaller than a constellation) is the Plough. Its stars, along with every visible star in the sky are slowly, imperceptibly, changing their relative positions. Civilisations living on Earth in a hundred thousand years will have to become familiar with a new set of constellations, and if we could travel forward in time then we too would not be able to recognise any constellations gracing the future skies.

**Constellations like Ursa Major will not appear
the same in a hundred thousand years.**

FACT 85

VY Canis Majoris is the largest known star in the Universe

Stars are big, very big. The Sun is an average star but even so, at 1.39 million kilometres across it is large enough that you could fit around a million planet Earths inside. Yet there are stars that dwarf even the Sun. The easily visible red giant star Betelgeuse in Orion is 965 million kilometres across, making its diameter 700 times that of the Sun. Even Betelgeuse is dwarfed by the hypergiant class of stars: these beasts of the stellar kingdom are goliaths. The largest known hypergiant is a star known as VY Canis Majoris. It has a diameter 1.98 billion kilometres across and if it were in the Sun's position, its outer layers would extend out to the orbit of Saturn.

VY Canis Majoris imaged by the Hubble Space Telescope. (NASA)

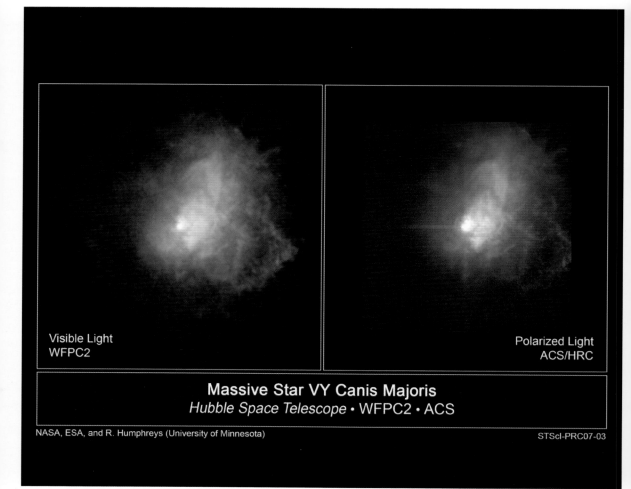

Visible Light
WFPC2

Polarized Light
ACS/HRC

Massive Star VY Canis Majoris
Hubble Space Telescope • WFPC2 • ACS

NASA, ESA, and R. Humphreys (University of Minnesota)

STScI-PRC07-03

VY Canis Majoris gets its name based on its variable nature. All variable stars have two letters followed by the Latin version of the constellation within which it is located. VY Canis Majoris designates the star as the 43rd variable star in the constellation Canis Major. It is a 9th magnitude star, which means it is a little over 16 times fainter than the faintest stars visible to the naked eye; in other words, you need a telescope to see it. Even though it is the largest known star in the Universe and a very luminous star at that, it appears so faint in our sky because it lies 3,820 light years away.

The true nature of this star has been the cause of quite some conjecture although it is now generally agreed that the star is embedded in a dense nebula which has several knots of nebulosity. Its variable nature was first suggested in 1931 and it was given its variable designation in 1939. It varies over a period of 956 days from its faintest at magnitude 9.6 to its brightest at 6.5, bringing it almost within the grasp of naked eye observation. Its variability comes from the gently pulsating nature of its growing and shrinking in size over the 965 day period, and it is this which causes its brightness in our sky to change.

There are no companion stars so estimating the mass is very difficult. Studying its other properties such as temperature (about 4,000 degrees compared to the Sun's temperature of 5,700 degrees) and luminosity enable a very rough estimate to be reached for its mass, which is thought to have started at around 30 times the mass of the Sun. Due to its immense size, its gravity is unable to hold on to the outer layers and it loses an estimated 0.0006 solar masses of material every year through a strong stellar wind.

FACT 86

Jupiter is a failed star

Jupiter is the largest planet in the Solar System, and at 139,822 kilometres across, 1,300 Earths would fit comfortably inside. Even though it is so much larger than Earth, it is only 318 times more massive, because it is made of gas and not rock. The composition of the gas that Jupiter is made of is very similar to the composition of the Sun, mainly hydrogen and helium with a few other elements mixed in for good measure. If Jupiter had been more massive (and estimates suggest it needs to be around 80 times more massive) then the conditions in the core may have been sufficient for hydrogen fusion to occur and for Jupiter to become a star.

Anyone that has taken a look at Jupiter through a telescope will notice that it looks quite different from the Sun. Of course you must never look at the Sun through a telescope without proper filtration because it can and will lead to blindness. Comparing the appearance of the Sun and Jupiter reveals some quite obvious difference: Jupiter is not giving off light and the Sun is a lot less colourful than the banded atmosphere of Jupiter. As far as the lack of light is concerned, it is the process of fusion in the core of the Sun which leads to the formation and emission of light. This fusion process, as we have seen, does not occur inside Jupiter.

Not only is the pressure lower inside Jupiter but the temperature too is far more moderate. The Sun clocks in at a little under 6,000 degrees but Jupiter is a much more chilly 145 degrees! This lower temperature means that hydrogen can remain bound to other atoms to form molecules such as ammonia and methane, giving Jupiter the colours we can see.

The key difference between the two is that Jupiter orbits something far more massive than itself and far more dominant whilst the Sun clearly is the powerhouse of the Solar System. When defining planets the IAU talks of a planet as something that orbits a star. Clearly therefore, this statement alone keeps Jupiter well and truly in its place as a planet. However there is a type of 'star' known as a brown dwarf. These 'star like' objects are failed stars much like Jupiter, the only difference between them is that brown dwarfs drift around the Galaxy unaccompanied whereas Jupiter is a member of the family we call the Solar System. If Jupiter were drifting in space then it too would be classed as a brown dwarf, a star-like object destined to drift among the stars without ever actually becoming one.

Jupiter imaged by the Hubble Space Telescope.
(NASA/ESA Hubble Space Telescope OPAL programme)

FACT 87

Dung beetles use the Milky Way to navigate

The behaviour of dung beetles seems really quite bizarre at first sight, they are attracted to great big piles of poo, then fashion some of it into nicely rounded balls and scurry off with them in a straight line away from the pile as fast as they can. The first question is why? As unpleasant as it sounds, dung beetles eat dung, so crafting them into spheres makes them easier to transport. Having worked hard to sculpt their next meal the last thing any dung beetle is going to want is for their dung ball to be stolen by another, so off they scurry to bury it far away.

This behaviour has been relatively easy to observe but what has perplexed scientists was exactly how they managed to roll the balls in a straight line! The terrain upon which this all takes place is usually pretty rough so managing to pick out a straight line is nothing more than incredible.

It has been known for some time that many insects, including our dung-loving friends, can see patterns of polarised light from the Sun and these are used by the beetles to navigate – but they can do it at night too. The fainter polarised light from the Moon can also be used but surprisingly the beetles can even maintain their straight lines on moonless nights. Researchers concluded that they must be using the stars to navigate. More specifically they thought, perhaps it was the Milky Way. To test their hypothesis the researchers setup a pile of dung in a planetarium and the beetles performed perfectly well under different sky arrangements. To test if it was the Milky Way the team removed all stars from the sky simulation except the band of light from the millions of stars in our Galaxy and sure enough the beetles were still able to travel in straight lines.

One final bit of confirmation was needed. They fitted the beetles with cardboard caps to block their view of the sky. As the researchers expected, the beetles were hopelessly lost and aimlessly wandered around in anything but straight lines, proving that dung beetles rely on the Sun, Moon and Milky Way to travel in straight lines.

Milky Way image captured from Norfolk, UK.
(Mark Thompson)

FACT 88

Jupiter is the Solar System's vacuum cleaner

The history of Earth would be very different if it were not for Jupiter, the largest planet in the Solar System. Since the formation of the Sun and its family of planets 4.6 billion years ago our region of the Galaxy has been swarming with chunks of rock. Their sizes range from sand grain size to big chunks many metres across.

Shoemaker–Levy 9 impacts on Jupiter, captured by the Hubble Space Telescope. (NASA)

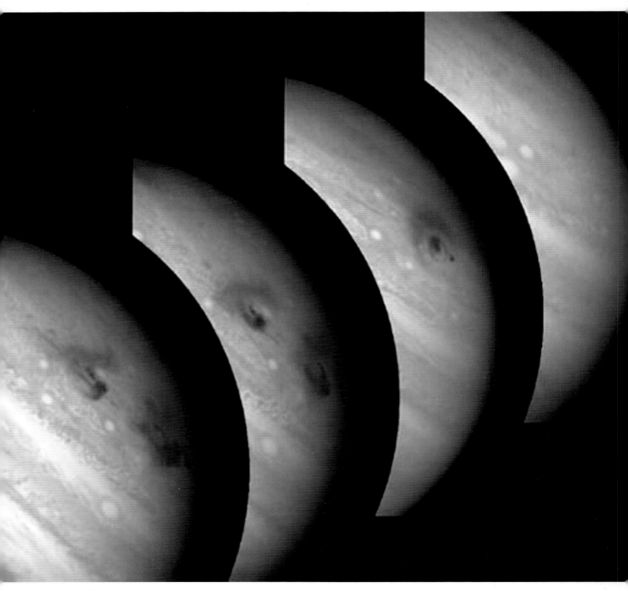

There is plenty of evidence that the Earth has been struck by pieces of space rock: Meteor Crater in Arizona, the Gulf of Mexico and Chicxulub crater to name just three of the most well known. Indeed the latter is the site believed to be the impact from the rock that wiped out the dinosaurs. But Jupiter, with its massive gravitational pull tends to protect us from many would-be intruders.

One such event occurred in July 1994 when the remnants of Comet Shoemaker–Levy 9 finally met its demise in the atmosphere of Jupiter. The comet had been discovered by Carolyn and Eugene Shoemaker and David Levy just a year earlier in March 1993 and is believed to have wandered too close to Jupiter around 30 years previously. The mighty pull from the gravity of Jupiter captured Shoemaker–Levy 9, swinging it into a spiralling orbit.

When the comet was first seen it became immediately apparent that this was no ordinary comet. A previous orbit around Jupiter had taken it too close and the immense tidal forces ripped the comet apart, to be precise, into 21 fragments. The fragments were of various sizes but the largest was approximately 2 kilometres across and each one, slowly and inescapably drifted towards Jupiter. At a speed of approximately 216,000 kilometres per hour, the fragments met their doom when they 'crashed' into Jupiter between 16 and 22 July 1994. The actual impact events took place just around Jupiter's limb so whilst the moment of impact could not be seen, the effects could definitely be observed, even with amateur telescopes.

The gas nature of Jupiter meant that the impact was not like a rock crashing into the solid surface of the Earth. Instead, when the fragments entered the Jovian atmosphere, they reached temperatures in excess of 23,700 degrees. As they plummeted deeper into the atmosphere the comet fragments were completely destroyed but debris could clearly be seen from Earth as dark spots slowly emerging around the limb of the planet. The spots ranged in size but were estimated to be several thousand kilometres.

Of course, had Shoemaker–Levy 9 wandered too close to Earth things would have been different. It may not have been caught by our gravity, and would have been less likely to have been broken up into fragments, but perhaps it could have crashed into Earth as one great big chunk, causing global devastation. Some estimates suggest that if Jupiter was not there, then the number of impacts suffered by Earth could be as much as 10,000 times greater.

FACT 89

Polaris is the north pole star but in 12,000 years it will be replaced by Vega

For thousands of years humans have been able to navigate their way around the northern hemisphere of Earth by identifying Polaris, the north pole star. The simple act of measuring its altitude or height above the horizon can quickly tell you your latitude. Yet things are changing and the sky too is changing, so much so that in 12,000 years, Polaris will be of less use to you and measurements of its altitude will not tell you where you are.

Polaris is otherwise known as Alpha Ursae Minoris, which designates it as the brightest star in the constellation Ursa Minor. It is indeed the brightest star but its neighbouring stars are faint. It is quite easy to miss Polaris in the sky, but at a distance of 433 light years, telescopic studies reveal that it is a multiple star system that comprises a main yellow supergiant star with two fainter companions.

The importance of Polaris is purely chance. The Earth is spinning on an axis that completes one revolution in just under 24 hours. This is what gives us the day. If you imagine all the stars in the sky fixed to a clear glass sphere that surrounds the Earth then it is easy to imagine extending the axis out onto the sphere. This gives us the so called north pole on Earth and directly above in the sky, the north celestial pole. It just so happens that Polaris is at this point in the sky! As the Earth spins it causes all the stars to appear to rotate around us but because Polaris is so close to the axis it barely moves in our sky.

The direction that the axis points changes over time. In the same way that a spinning top wobbles as it slows down, so the axis of the Earth wobbles, and slowly, over thousands of years, scribes out a circle in the sky. Currently it is pointing at Polaris but in 12,000 years it will have slowly moved and will be pointing towards Vega in Lyra. Some might say that Vega is a much more worthy star to become the pole star because it is much brighter. Unlike Polaris, which is pretty inconspicuous, Vega is the fifth brightest star in the sky. Vega is the brightest star in Lyra so its designation is Alpha Lyrae. It is much closer to Polaris at just 25 light years and shines with a bright blue-white colour.

Polaris can be found using the pointer stars in the Plough. (Mark Thompson)

FACT 90

The night sky in a globular cluster would be glittering with thousands of bright stars

The night sky has a few bright stars. Sirius and Vega are great examples but they are often outshone by the planets Venus and Jupiter. In general, the stars are nicely visible but they could not be classed as spectacular. In our region of the Galaxy the stars are quite sparsely populated; even the nearest star is still 4.3 light years away and whilst the actual brightness of a star is a major factor of how bright it appears in the sky, so is its proximity to us. Having other stars too close is likely to prohibit the evolution of a stable planetary environment upon which life could form but if you happened upon a planet in a globular cluster then your sky would be glittering with thousands of bright stars.

Globular clusters are, as their name suggests, broadly spherical in shape. They can be seen in our night sky as faint fuzzy blobs, and in the nearer ones, we can even see individual stars. There are about 150 known globular clusters around our Galaxy but larger galaxies can have more. Clusters can vary in the number of stars they host, from a few hundred thousand to the largest hosting several million! Omega Cenaturi is the largest known globular cluster of the Milky Way and is found 15,800 light years away in the constellation of Centaurus. The 10 million stars believed to be hosted in the cluster are enclosed in a region of space just 150 light years in diameter that can be compared to an estimated 8,000 stars within our region of the Milky Way!

The density of stars inside a globular clusters is pretty incredible and in most cases, the distances between stars is fractions of a light year rather than several light years. Taking a cube sized chunk of space from a globular cluster with sides equal to the distance between us and our nearest star would net not one star but in the region of 1,000 stars.

With such proximity between the stars, the intensity of radiation and the tidal effects from their constantly changing positions would mean planets stand very little chances of forming. If one did happen to form then there is a very good chance it would not enjoy much stability before being flung out of the cluster. If perchance you were able to gaze upon an alien sky from a planet in a globular cluster then the number of stars close to you in space would present you with a sky full of thousands of glittering stars of the brightness of Venus. Their combined light would allow you to walk around easily and see what you are doing even without any moonlight.

FACT 91

Eratosthenes measured the circumference of Earth in 240bc

Science is all about observation, and one Greek astronomer, Eratosthenes, made an observation around 240bc that enabled him to calculate (with quite an impressive degree of accuracy) the circumference of the Earth. He lived in Alexandria, which was a city in northern Egypt, near where the Nile meets the Mediterranean. He had travelled considerably and noticed that at a certain time of day on a particular day of the year, the Sun shone to the bottom of a well in the town of Syene in southern Egypt. The time was noon and the day of the year was the summer solstice. This told Eratosthenes that the Sun was directly overhead at that time.

The observation that Eratosthenes made was crucial to his calculations but the next step was the insightful realisation that back home in Alexandria further north, the Sun was never overhead. He concluded that if he measured how far away from overhead the Sun was, then he could use simple geometry to calculate the circumference of the Earth. To work out the angular distance the Sun was from overhead at the same time, he measured the angle formed from a shadow that was cast by a vertical tower. He measured an angle of 7.2 degrees and hypothesised that this angle would be the same as the angle between two imaginary lines from Alexandria and Syene to the centre of the Earth. Knowing that there were 360 degrees in a circle and that 7.2 degrees divides nicely into 360 degrees by 50, if he could find the actual distance between Alexandria and Syene then he need to multiply the result by 50 and he would have the circumference of the Earth!

Records get a bit sketchy here though. The distance between the two places was recorded as 5,000 stadia, which meant his calculation of the circumference of Earth was 250,000 stadia. Now not many people are familiar with the measurement of a stadion (singular of stadia) as it was not used widely. The scale has its origins in the size of a sports stadium, hence its name, but of everyone who used the stadion, they did not all agree on its length.

We now have a very accurate value for the circumference of the Earth but it is not perfectly spherical. The circumference around the Equator is 40,075 kilometres but around the poles it is a little less, at 40,008 kilometres. Even so, depending on the value for the stadion used, then Eratosthenes was very close, within 1% of today's value, but if we use the upper end of the value then he was only 15% of today's value. Either way, given that he made his measurements over 2,000 years ago it is still a very impressive feat.

FACT 92

A star called Lucy is a large cosmic diamond

Lucy is a star that was discovered in the constellation Centaurus but its official name is BPM37093. It also goes by the name of V866 Cen, designating it as the 886th variable star in Centaurus. It lies about 50 light years away so in cosmological terms is right on our doorstep, but even though it is so close it was only spotted for the first time in 2004 and cannot be seen without the aid of a telescope. What really excited astronomers though was that observations seem to reveal that 'Lucy' is a massive cosmic diamond, or more accurately, a 5 million trillion trillion pound chunk of crystalline carbon, that equates to a diamond of 10 billion trillion trillion carats! A carat is a measure of the weight of a diamond and to put that into context, a diamond that weighs the same as a paperclip would equate to 1 carat.

The crystalline carbon structure of the star led to its affectionate name of Lucy after the Beatles song Lucy in the Sky with Diamonds but in reality Lucy is a white dwarf star. These stellar corpses are what is left behind when a star much like the Sun dies. Throughout their life they will have fused hydrogen into helium and helium into carbon and oxygen, and as the star died most of its outer hydrogen and helium layers will have been lost to space. This is the fate that awaits our Sun, but the Sun is currently about 1.4 million kilometres across whereas Lucy is only 4,000 kilometres in diameter. Even though Lucy is smaller than Earth it still has approximately the same mass as the Sun – that is one dense object.

Lucy is known as a pulsating white dwarf. For most of its life, the force of gravity that tried to collapse the star was balanced by the thermonuclear force, the result of nuclear fusion. As relatively low mass stars like these die, they expand and lose some of their outer layers to space, exposing their core, which over time, cools down. This exposed and cooling core becomes the white dwarf, but many will begin to pulse when the surface temperature of the core drops to around 12,000 degrees. This is in comparison to the Sun's core temperature of approximately 15 million degrees. The pulsations of Lucy and similar types of stars are like seismic waves on Earth which can be used to probe the Earth's interior. Studying the pulsations of Lucy enables astronomers to analyse the inside of the star and determine that it is solid crystallised carbon.

Putting all of this together, we have a star which is the structure of diamond. The largest diamond here on Earth is known as the Golden Jubilee Diamond and comes in at a whacking great 545 carat, but if you have not heard of that then you may have heard of the Great Star of Africa which weighs in at 530 carats. Now imagine a diamond not weighing in at just over 530 carats but instead 10 billion trillion trillion carats! That is one big diamond!

FACT 93

Lunar eclipses cause massive temperature drops on the Moon

Lunar eclipses are beautiful events to witness. Unlike their solar counterpart lunar eclipses can be seen from anywhere on Earth where the Moon is visible. If a lunar eclipse is visible where you are then you might get to see a total eclipse where all of the Moon is eclipsed or you might see a partial eclipse where only part of the Moon is obscured.

The mechanics of a lunar eclipse are simple to understand. We see the Moon because it reflects sunlight, and if we somehow block sunlight from reaching the Moon then it will go dark. If you have ever seen a lunar eclipse you will know that they only occur at full moon, which takes place when the Earth lies between the Sun and Moon so we see the fully illuminated face of the Moon. It is also important to note at this point that the Moon goes around the Earth once every month. Indeed the word Moon has origins in the word 'month', but the plane of the Moon's orbit is tilted very slightly to the plane of the Earth's orbit around the Sun. This means that during most full moons the Moon is slightly above and sometimes slightly below the line joining the Sun and Earth. The three must be in perfect alignment for a total lunar eclipse, an alignment we call a syzygy.

If the three are in perfect alignment then the Moon will gracefully drift through the shadow cast by the Earth from the Sun. It is unusual for the Moon to completely vanish from view even with the Earth in the way, because the longer wavelength red sunlight gets bent through our atmosphere to gently illuminate the shadow region in a red glow. Quite often a total lunar eclipse will reveal a Moon glowing a beautiful red colour. Sometimes the Moon can skirt along the edge of the Earth's shadow to give us a partial lunar eclipse where only part of it goes dark.

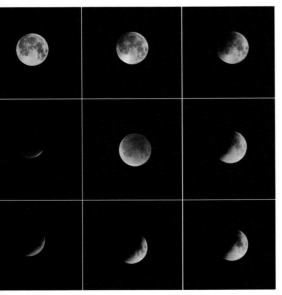

Lunar eclipses occur when the Earth is between the Sun and Moon. (Mark Thompson)

Think about a lunar eclipse from the point of view of the Moon though. As the Moon moves into the shadow of the Earth, the Moon will experience the shadow slowly slipping across the lunar surface. If you have ever been outside in the heat of the mid-day Sun you will know that standing in a shadow makes you feel suddenly quite a bit cooler. It is the same for the surface of the Moon during a lunar eclipse. During a lunar eclipse in 1971, temperatures were monitored at two Apollo landing sites, and as the shadow drifted over, a temperature drop of 307 degrees was recorded. With such a sudden change in temperature, the thermal shock can cause lunar rocks to crumble and allow gas to seep out from the interior of the Moon.

FACT 94

A teaspoon of neutron star material weighs 10 million tonnes

Atoms have a varied structure depending on the element they belong to but they all share common components. At their core they will be composed of any number of protons with a positive charge and neutrons with a neutral charge, and surrounding this nucleus can be a number of electrons with a negative charge. Hydrogen is the most basic atom and contains a single positive proton and a single negative electron, giving the atom a neutral charge.

When stars form they are mostly composed of hydrogen although there are heavier elements present in many new stars. Throughout a star's life they fuse hydrogen into successively heavier elements such as helium, carbon and silicon, but quite how far they go depends on their mass. Stars like our Sun will fuse hydrogen into helium and helium into carbon and oxygen, but for more massive stars, they will fuse carbon and oxygen into heavier elements. Stars that are between 10 and 29 times more massive than the Sun are able to synthesise iron in their core.

It is not possible to fuse iron and get energy out so the nuclear fusion in the core of the star stops and the star collapses. As the star collapses, the core is under immense pressure and the only thing that can stop its ultimate collapse into a black hole is the force binding the atoms together. The crushing pressure from the collapse causes protons and electrons to combine to form neutrons which release bursts of neutrinos out into the Universe. Neutrons though, are like engineering bricks, they can withstand more pressure than other bricks, enough to halt the collapse of a star.

The final result of this cataclysmic collapse is a neutron star, and current estimates suggest there are 100 million of them in the Milky Way alone. They are stars that have around 1.5–2 times the mass of the Sun yet occupy a space not much bigger than 20 kilometres in diameter. In fact, so much material has been crushed together that if you could get your hands on a super strengthened teaspoon and scoop up a good spoonful it would weigh in the region of 10 million tonnes. Due to the law of the conservation of momentum, the slowly rotating, collapsing star would have to maintain its momentum, so with its decreasing size, the rotation rate increases. It is not uncommon for neutron stars to rotate many hundreds even thousands of times per second. The dead, fast-spinning star is essentially one massive neutron formed when all of the empty space between the neutrons in the star's core simply gets squeezed out. Now that the core has no material in it to fuse, there is no more energy being created and the fast spinning star will slow and cool over millions of years, eventually dimming away into obscurity.

Isolated Neutron Star RX J185635-3754

Hubble Space Telescope • WFPC2

PRC97-32 • ST ScI OPO • September 24, 1997
F. Walter (State University of New York at Stony Brook) and NASA

Neutron star seen by the Hubble Space Telescope. Studies revealed the star was at 670 thousand degrees. (NASA/ESA)

FACT 95

The Pistol Star is 10 million times brighter than the Sun

Stars come in all sizes and brightnesses. Our Sun is pretty average as they go, not particularly large nor particularly bright. Of course its proximity to us means that it is particularly prominent in our lives, but push it out to the distance of a few light years and all of a sudden that bright thing in the sky that lights our day becomes pretty insignificant. There are stars out there that are much brighter than the Sun. You only have to look up in the sky to see bright stars and faint stars but that can be misleading. A bright star a long way away can look quite faint. Indeed one of the most luminous stars in the Milky Way is known as the Pistol Star, yet it is impossible to see visually, even with a telescope.

At a distance of approximately 25,000 light years from Earth and hidden from view by a massive molecular dust cloud is the Pistol Star. Even at that great distance, if it was not for the dust cloud, the star would be visible to the naked eye, but alas the cloud blocks its visible light so it must be studied in other wavelengths such as infra-red. It was discovered by the Hubble Space Telescope in the early 1990s using its Near Infrared Camera and Multi-Object Spectrometer.

The Pistol Star is a blue hypergiant star, which is quite rare. These are stars that begin life with around 40 times the mass of the Sun, and due to their extreme mass are unable to maintain much stability; instead they lose much of their initial mass through colossal outbursts of solar wind. It is thought that the Pistol Star, whose official designation is V4647 Sgr, lost almost 10 times the mass of the Sun in a massive outburst. Given the immense distance to the star, this outburst, whilst seen from Earth around 1500AD would actually have happened 25,000 years previously, around 23500BC. It is thought that this cloud is now visible as the Pistol Nebula, which enshrouds the star and hides it from our view. The whole complex gets its name from the shape of the nebula, which in low resolution photographs rather resembles a pistol!

It is difficult to probe the secrets of the star due to the obscuring cloud, and indeed estimates of its age, mass and luminosity are uncertain. It is thought, due to its mass, that it will evolve quickly and go hypernova in just a few million years but before it does it will continue to emit astronomical amounts of energy. Current estimates suggest it is 1.6 million times more luminous than the Sun and that it is emitting as much energy in 20 seconds as the Sun emits in one year!

Pistol Star and Nebula
Hubble Space Telescope • NICMOS

PRC97-33 • ST ScI OPO • October 8, 1997 • D. Figer (UCLA) and NASA

The Pistol Star is 10 million times brighter than the Sun. (NASA)

FACT 96

The atmosphere of the Sun is hotter than its visible surface

The inner planets of our Solar System: Mercury, Venus, Earth and Mars, are all rocky bodies that if you visited, you would find a solid surface upon which you could land and walk around. The environment may be quite hostile but still there would be solid ground upon which to plant your feet. The Sun is quite different for it is a giant ball of gas 1.4 million kilometres in diameter. Even with the gaseous nature of the Sun we can still divide it up into quite distinct regions from its core at the very centre through the radiative zone, the convective zone, the visible surface we call the photosphere, the chromosphere and the outer atmosphere called the corona.

The solar corona captured by the SOHO spacecraft. (NASA)

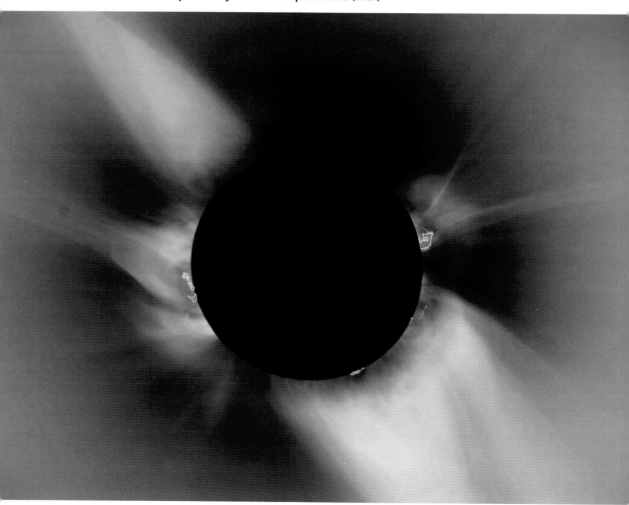

Nuclear fusion takes place in the core where the heat and light are generated which are so important to our survival. You might expect that as you travel further from the core of the Sun, the cooler it gets and this is generally what we see but one discovery has rather perplexed solar scientists. The corona which is the outermost region of the Sun is hotter than the photosphere ... 300 times hotter!

The photosphere is a 500 kilometre thick, visible layer of the Sun where the energy generated in the core is released as light. The temperature of the photosphere is around 6,000 degrees and it is surrounded by the much thinner and cooler chromosphere at around 4,500 degrees. Yet both of these layers are positively chilly when compared to the outer corona, which reaches temperatures of a few million degrees! Such a sharp increase in temperature between the photosphere and corona has been plaguing scientists for years. One possible explanation is the violent solar flares that erupt from the Sun during moments of activity.

As the Sun rotates, the equatorial regions move faster than the polar regions causing magnetic field lines to get dragged around, they get wound up tighter and tighter until they snap, bursting through the surface, creating sunspots. They also seem to be accompanied by short eruptions of high energy radiation, which can often be observed as bright flashes of light. These flares, or their smaller nanoflare cousins may well be the mechanism that is transferring heat into the corona.

Nanoflares are tiny in comparison to regular flares and only have a billionth of their energy but even so, it is estimated that the nanoflares can still reach temperatures in the region of 10 million degrees. One nanoflare is thought to carry the equivalent energy of a 10 megaton hydrogen bomb and these going off randomly around the Sun every few seconds might be enough to transfer energy through to the outer layers of the Sun's corona. They may be accompanied by giant solar twisters, the result of twisted magnetic field lines and solar material. It is thought that these (and there may be of the order of 10,000 of them at any one time!) along with the nanoflares will be giving the corona the extra thermal boost it needs.

FACT 97

Some rocket engines produce enough thrust to lift a sheet of A4 paper!

Think of a rocket and most people have visions of something along the lines of the mighty Saturn V that produced 7.5 million pounds of thrust and sent Neil Armstrong, Buzz Aldrin and Michael Collins towards the Moon. The measure of the efficiency of a rocket engine is defined by its specific impulse, which articulates the change in momentum created by a given amount of propellant. In the case of the Saturn V rocket, this was 263 seconds. The technology behind the Saturn V was liquid propellant: take some liquid hydrogen and liquid oxygen, pump them into the rocket

A multi-thruster ION engine. (NASA)

chamber, BOOM and you have LOTS of energy which is forced out of the rocket nozzle, pushing the rocket forwards (or upwards, depending on how you look at it).

In order to get a rocket (and cargo) away from Earth, the brute force of the liquid-fuel rocket engine is perfect. The gravity of Earth is such that any rocket wanting to escape needs to travel at 11.2 kilometres per second. Once in space and away from the confines of gravity, rockets do not need to pack quite so much of a punch. In space, brute force is not quite so crucial, so instead, rocket designers can rely on a more efficient design.

The ion engine is a great example where efficiency outweighs brute force, and they are powered by gas. Atoms contain protons and neutrons in their core and they are surrounded by a number of electrons, and in a gas, atoms move around with weak bonds between them. When a gas atom becomes ionised, it has either gained or lost an electron so that the atom is no longer neutral but is either positively or negatively charged. These ions are used as propellant for ion rocket engines.

Most ion engines bombard atoms of usually xenon gas with high energy electrons. This knocks electrons out of the Xenon which results in a mixture of positive xenon gas atoms and freely roaming negative electrons. Xenon is generally used because it has a high atomic mass (equal to the number of protons and neutrons in the nucleus) and can therefore produce higher amounts of thrust than a gas with a lower atomic mass. At the rear of the rocket chamber is a grid known as the accelerator grid which is negatively charged. The positive ions are accelerated towards the accelerator grid, reaching speeds of around 144,000 kilometres per hour where they are ejected out of the engine, generating tiny amounts of thrust to propel the rocket forward. The beam of ions being ejected from the engine are neutralised by a stream of electrons bombarding it.

The ion engine has a much higher specific impulse than the Saturn V rocket, for example the NASA Evolutionary Xenon Thruster engine has a specific impulse of 4,190 seconds as against Saturn V's 263 seconds). This sounds very impressive but it equates to tiny levels of thrust, just enough in fact, to propel a sheet of A4 paper! In deep space, high thrust is not needed. As long as you are not in a rush to get anywhere, then low, consistent thrust is far more efficient than high levels of thrust. Saturn V sent *Apollo 11* to the Moon in four days, consuming 40,000 pounds of fuel every second during launch, whereas *Smart-1* used an ion engine and took a more leisurely pace, arriving 1 year, 1 month and 2 weeks after launch but only consuming around 170 pounds of fuel for the whole journey!

The Sun is a very faint star!

Look up at the night sky and it is clear for anyone to see that the stars are all of different brightness. There is no upper limit to the brightest stars that we can see with the naked eye but there is a lower limit. The Sun is the brightest thing we can see in the sky but it would be quite wrong to think that it is the brightest of all objects in the Universe. Along with the amount of light being kicked out by a star, its brightness in our sky is also determined by its distance. When we take into account the relative distance of the Sun and stars, it turns out the Sun is not so bright after all.

Hipparchus was a Greek astronomer who devised a scale to measure stellar brightness and we still use it today. It is known as the magnitude scale and when it refers to the brightness of objects as they appear in the sky, we call it the apparent magnitude scale. The brightest stars visible in the sky (ignoring the sun) are classed as 1st magnitude and the dimmest stars visible are 6th magnitude, therefore a higher number actually means a fainter star. The difference in brightness between each magnitude is 2.5 times so a 2nd magnitude star is 2.5 times fainter than a 1st magnitude star and a 6th magnitude star is 100 times fainter than a 1st magnitude star. Since Hipparchus devised the scale it has been extended to include all objects in the sky, including those that we cannot see with the naked eye. On this new extended scale, the full Moon is magnitude -12.6, Venus has a magnitude of -4.6, stars only visible through telescopes have a magnitude higher than 6 and it might not surprise you that the brightest object in the sky, the Sun, has an apparent magnitude of -26.74.

Apparent magnitude, as its name suggests relates only to the apparent brightness of an object and not its real brightness. If we wish to compare the real brightness of objects then we need to remove the distance factor which effects the brightness of objects in the sky. Enter the absolute magnitude scale. This evolution of the apparent magnitude scaled compares the brightness of objects based on how they would appear at a distance of 32.6 light years.

Taking the Sun and pushing it 32.6 light years away is going to make it fainter but then other stars, such as the Pistol Star, which is one of the most luminous stars in our Galaxy, is 25,000 light years away so bringing that closer is going to make it appear brighter. Using this new absolute magnitude scale we can compare the actual or instrinsic brightness of objects. Sirius for example, which is the brightest star in the sky, has an apparent magnitude of -1.4 but an absolute magnitude of 1.4; the Pistol Star has an absolute magnitude of -10.7 (so it would easily outshine Venus at its brightest); and the Sun, which has an apparent magnitude of -26.74 has an absolute magnitude of 4.83 so it would only just be visible to the naked eye.

FACT 99

Astronomers have their very own tape measures

Think about the size of your lounge and you will probably think of it in terms of metres. Turn your thoughts to the distance to work, or to your local pub and you will more than likely think of it in terms of miles or kilometres. Astronomers can only use these familiar terms when talking about small measurements in the Universe, for example the diameter of a planet or the distance between the Earth and Moon. As the size and distances increase, we need to use different tape measures to define them. If we do not, then the numbers start to get unwieldy. For example the diameter of the Milky Way galaxy is 1 billion billion kilometres and as you will agree, getting your head around numbers like that is pretty tricky. Instead, astronomers have their own scales to make things easier.

In the Solar System, the distances are occasionally measured in kilometres. For example the average distance between the Earth and Sun is 150 million kilometres, but even these numbers can get quite big. Instead, we use a scale based on the average distance between the Earth and Sun and call it 1 astronomical unit. On this scale, Saturn is 9.6 astronomical units (or 9.6 times the distance between the Earth and Sun) from the Sun, easier to deal with than 1.43 billion kilometres.

Move out among the stars and even the astronomical unit becomes too small to deal with the distances. So we turn to the light year. Its name suggests it has something to do with time and it does, but not in the sense you might think. Light travels fast, in just one second it can traverse 299,792 kilometres; in a year it can travel 9.5 trillion kilometres. A light year is the distance light can travel in one year so an object that lies 1 light year away means it is 9.5 trillion kilometres away. This also equates to 63,242 astronomical units.

The astronomical unit and the light year are perhaps the most commonly used distance scales in astronomy, but even using them, the most distant objects in the Universe are around 13 billion light years away, and as you can perhaps appreciate, these numbers are still large and unwieldy. There is one more distance measure used to make things a little easier and its called the parsec but its origin is a little more complex to understand.

There are 360 degrees in a circle, but each degree can be divided into 60 equal chunks to give a minute of arc, and each minute of arc can be divided further into 60 chunks giving us seconds or arc. An arc second is a tiny angular measure and there are 3,600 in one circle. As the Earth moves around the Sun the nearby stars seem to shift their position by a very small amount. When a star is at such a distance that this apparent shift equates to 1 second of arc, that star is said to be at 1 parsec, and it equates to 3.26 light years. The parsec can be further scaled up to the kiloparsec which is a thousand parsecs, or the megaparsec which is 1 million parsecs. Using this, the largest of all the distance scales, the most distant galaxies are 3,985 megaparsecs away, much easier than 100 billion trillion kilometres!

FACT 100

A Martian meteorite has the fossilised building blocks of life inside it

On 27 December 1984, a team of American meteorite hunters picked up a meteorite from a region in Antarctica known as Alan Hills. The meteorite carries the designation ALH84001 and is one of a number of meteorites collected from the area due to the ease with which they are found. On first inspection, a meteorite looks just like a terrestrial rock, so much so that if you had one in your garden, you would be hard pushed to pick it out among the hundreds of other rocks. Look in Antarctica though, and any meteorite sitting on the icy flats looks pretty obvious. ALH84001 was no ordinary meteorite though: it had travelled all the way from Mars and locked up inside were the fossilised building blocks of life!

ALH84001 is thought to have crystalised from molten rock a little over 4 billion years ago on Mars and lay undisturbed for millions of years. A chance meteoric impact on the red planet dislodged the rock 17 million years ago, blasting it into space. There it stayed, drifting around the Solar System until around 13,000 years ago when it fell to Earth. There it sat, on the Antarctica ice until it was collected for study in 1984.

It was easy for the researchers to confirm that the meteorite had indeed come

The surface of Mars, origin of the ALH84001 meteorite.

from Mars, since they found tiny little pockets of gas trapped in the rock from when it crystalised during formation. These gas pockets were carefully analysed and revealed the same chemical composition as the Martian atmosphere as revealed by the many spacecraft that have already sampled its atmosphere.

Martian meteorites were nothing particularly new, as others have been previously identified, but what was much more of a surprise were the organic compounds that were found inside. Some claimed that since the rock had been sat in Antarctica for around 13,000 years then there was a good chance it had been contaminated from Earth based compounds. The distribution of the compounds makes this quite unlikely however since there are more towards the centre than there are towards the outside. This is more likely to be opposite had Earth compounds contaminated the meteorite.

The likelihood of the meteorite having come from Mars was really quite high, as was the origin of the organic compounds which were very likely to have also come from Mars. But what are they? Scientists agree that they are not 'life'. The strings of organic compounds in the ALH84001 meteorite are now thought to be the fossilised building blocks of life, which in itself is a remarkable discovery. That some of the building blocks for life started to try and establish themselves on Mars millions of years ago suggests that the conditions for the evolution of life on Earth may not be that unique after all.

FACT 101

There are alien lakes on Saturn's moon Titan

The European Space Agency launched the *Cassini–Huygens* Mission to Saturn in October 1997. After seven years coupled together, the *Huygens* probe detached from *Cassini* and began its descent down to the surface of Titan, Saturn's largest moon, and landed on 14 January 2005. The probe weighed in at 318 kilogrammes, its heat shield was 2.7 metres in diameter and inside, the probe itself was smaller at 1.3 metres across. Its descent through the atmosphere was slowed by parachute as atmospheric measurements were taken until it finally landed on the surface, revealing wonderfully mystical landscapes.

Huygens' landing site had been previously identified, and images taken by *Cassini* when it was still just over 1,200 kilometres away revealed what looked like a shoreline. The probe had been designed to account for as many eventualities as possible so it had been designed to be able to withstand a landing on solid ground or in liquid and with enough power to transmit telemetry back to Cassini and on to Earth for several minutes.

The successful landing of *Huygens* on Titan occurred at 12:43 on 14 January 2005. With the parachute braking technique the final impact would be equivalent to dropping it on Earth from a height of about one metre. Having bounced and slid around a little, the probe settled down, started capturing images of its surroundings and sending them back to mission control. The landing site showed what looked like pebbles of water ice scattered around as you might expect pebbles of stone to be scattered around a dried up river bed here on Earth. The first images from the surface of Titan were incredible but they were nothing compared to the images snapped by *Huygens* on the way down.

The descent pictures showed dark features, some of which seemed to have jagged lines sticking out of them like river tributaries on Earth. These are thought to be channels of liquid traversing the lighter coloured terrain before meeting with the sea. Do not be misled though, these bodies of liquid on the surface of Titan are not watery pools you could plunge in to; they are quite different. The temperature at the surface was recorded to be -179 degrees, far too cold for liquid water but not too cold for liquid methane. The bodies of liquid hinted at by *Huygens* in its descent images and revealed in incredible clarity by *Cassini* are bodies of liquid methane.

It seems then, that if you managed to land your spacecraft on the shore of one of these alien lakes then you would have a lake of methane stretching out ahead of you. Further radar observations by orbiting *Cassini* revealed that these beautiful lakes were mirror smooth; not a wave, nor disturbance seemed to break the surface. Wind on Earth causes waves on our oceans so if the lakes of methane show not even a ripple, perhaps this can be taken that the weather on Titan is calm.

Composite image showing the dark alien lakes on Titan. (NASA)